T0300850

Analog
and
Pulse Circuits

Analog
and
Pulse Circuits

Dr. Dayaydi Lakshmaiah
Professor
Department of ECE
SIIET, Hyderabad

Dr. C. B. Ramarao
Professor
Department of ECE
NIT, Warangal

Dr. K. Kishan Rao
Professor
Department of ECE & Ex Director
NIT, Warangal

BSP **BS Publications**
A unit of **BSP Books Pvt. Ltd.**
4-4-309/316, Giriraj Lane, Sultan Bazar,
Hyderabad - 500 095

First published 2022
by CRC Press
4 Park Square, Milton Park, Abingdon, Oxon, OX14 4RN

and by CRC Press
6000 Broken Sound Parkway NW, Suite 300, Boca Raton, FL 33487-2742

© 2022 BS Publications

CRC Press is an imprint of Informa UK Limited

Print edition not for sale in South Asia (India, Sri Lanka, Nepal, Bangladesh, Pakistan or Bhutan).

British Library Cataloguing-in-Publication Data
A catalogue record for this book is available from the British Library

Library of Congress Cataloging-in-Publication Data
A catalog record has been requested

ISBN: 978-1-032-22869-3 (hbk)
ISBN: 978-1-003-27458-2 (ebk)

DOI: 10.1201/9781003274582

Dedicated to

My Parents
D. Bala Naganna, D.Sirovanamma
and

My wife
D. Benny

Our child
D. Venkata Siddhartha,
D. Sai Harshitha

... Dr. D. *Lakshmaiah*

Preface

After nearly thirty five years of experience in classroom .The purpose of writing this book is to present undergraduate students the wide-ranging concepts of Analog and Pulse Circuits in a reader-friendly style. I do not fully subscribe to the oft-repeated opinion that the study of analog circuits designed with diodes, transistors, and other discrete components is less rewarding in an age where digital systems like microprocessors, microcontrollers, personal computers, and embedded systems have a decisive edge in electronic circuit design. While acknowledging the strong impact of the digital world and personal computers in all walks of our present-day life, it is hasty and unwise to deny the core subjects the importance they deserve. It is the responsibility of the academic fraternity to motivate budding students of electronics to learn and strengthen their roots in electronics. Undoubtedly, Pulse and Digital Circuits is one such core subject that an electronic engineer has to master in order to build his creative genius in the analysis and design of both analog and digital systems. The very fact that a large number of quality books being published every year on analog circuits offer sample proof that these difficult subjects are going to be in need for many years to come. It has to be admitted that the study of subjects of this nature demand more patience and perseverance. The best way to learn these subjects is obviously to study them under the expert guidance of good teachers. A good teacher is worth a million books. I had the good fortune of studying this subject from highly dedicated teachers, and I wish to pass on the benefits I derived to the present-day student community.

The importance of electronics is well known in various fields of engineering. It's before necessary for an engineer to know the fundamentals of pulse and digital circuits the book covers the entire syllabus of subject. pulse and digital circuits, I have strived to develop this compensative text on pulse circuitry in order to provide students with a solid ground in the foundation of analysis and design of pulse and digital circuits.

The book uses the plain and simple language to explain the subject the book prepares carefully the background of each topic giving suitable practical illustrations. Full detailed of each step is given. It is easy for the students to understand the complicated derivations.

Chapter 1: When non sinusoidal signal are transmitted through a linear network the shape of the waveform undergoes a change this process is called linear wave shaping which is discussed in chapter -1 signals like step, ramp, exponential, pulse and square waves change shape when passed through low pass and high pass RC circuit. RC low pass and high pass, RLC circuits and attenuators are describes in this chapter.

Chapter 2: In communication systems it is required to remove a part of waveform above or below some reference level. This process is called clipping. DC level is required to be added to waveform to fix the top or bottom of the waveform at same reference level the process is called clamping in chapter -2 deal with various one level and two level , series and shunt and dampening .the analysis of clipping and clamping circuit with examples.

Chapter 3: The switching characteristics of junction diodes and transistors as required for clear understanding of pulse and digital circuits are covered in chapter -3. On switching characteristics of devices introduces the charge control model of a diode for the purpose of studying its switching behaviour. After studying the law of the junction, we define the forward recovery time and reverse recovery time of a diode. The two major breakdown mechanisms are also briefly studied here as they are related to the switching behaviour. Later the bipolar junction transistor is studied as a switch and delay time, rise time, storage time, and fall time are defined. The chapter also deals with transistor breakdown voltages and temperature variation of saturation values. Other miscellaneous topics like collector-catching diodes, transistor switch with inductive load, damping diodes, transistor switch with capacitive load, and NMOS, PMOS and CMOS switches are also studied

Chapter 4: Memory is the basic requirement for all computers. The basic memory element is a flip-flop, i.e. The bistable multivibrator the monostable multivibrator is the basic gating circuit. The astable multivibrator is used as a master oscillator, and the Schmitt trigger as a basic voltage comparator. the various types of multivibrators –(collector coupled fixed bias type and self bias type, emitter coupled type, i.e. Schmitt trigger), monostable (collector coupled , emitter coupled) and Astable are discussed in chapter -4. The analysis and design of all these multivibrators are also illustrated with examples in this chapter.

Chapter 5: Time base generators are essential for display of signals on screen. Time base generators may be voltage time-base generators or current time-base generators. Various methods of generating time baser waveforms, transistor voltage time-base generators –miller and bootstrap type and current time base generators are discussed in chapter -5 the analyses and design of time base generators are also lustrated with examples.

Chapter 6: Logic gates are the fundamental building blocks of any digital system the basic gates AND, OR, and NOT, the universal gates-NAND and NOR, the derived gates- XOR and X-NOR and the realization of logic gates using diodes and transistors are discussed in chapter -6 inhibit circuits and pulsed operation of logic gates are also discussed in this chapter.

Chapter 7: When signals are to be transmitted only for specified intervals of time and are to be blocked during other intervals of time, we require sampling gates which may be unidirectional or bidirectional unidirectional diode sampling gates, bidirectional diode sampling gates bidirectional transistor sampling gates, two-diode gates, four-diode gates and six-diode gates and also applications of sampling are discussed in chapter -7.

Contents

CHAPTER 6

CHAPTER 7

1

Linear Wave Shaping

.1 INTRODUCTION

Let us consider a transmission network consisting of linear elements. Sinusoidal signal is applied to a network, the output signal is sinusoidal in the steady state conditions. The influence of the network circuit on the signal may be completely specified by the ratio of output to input amplitude and phase angle between output and input waveform. No other periodic waveform preserves its shape. Generally when transmitted through a linear network the output signal may have a little resemblance to the input signal.

"The process whereby the shapes of non sinusoidal signals are shaped by passing the signal through the linear network is called linear wave shaping".

.2 HIGH PASS RC CIRCUIT

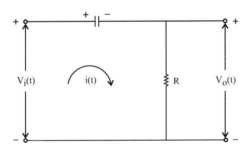

FIGURE 1.1 High pass RC circuit

The high pass RC circuit is shown in Fig.1.1. The input is denoted by $V_i(t)$, and the output as $V_o(t)$, 'a' is the charge of the capacitor.

At zero frequency the capacitor has infinite reactance and hence open circuited. Therefore, the capacitor blocks the dc signal not allowing it to reach output. Hence the capacitor is called blocking capacitor. The coupling circuit provides dc isolator between input and output.

Since the reactance of the capacitor decreases with increasing frequency the end output increases.

Thus the circuit abstracts the low-frequency and it allows the high frequency to reach the output. Hence this circuit is called high pass RC circuit.

1.3   SINUSOIDAL INPUT

The sinusoidal input V_i (t) is mathematically defined as Vi (t) $= V_m$ sin wt

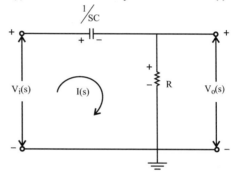

FIGURE 1.2   Laplace Network of high passe RC circuit

In the analysis of Network to sinusoidal input is obtained using Laplace transform as shown in Figure 1.2 applying KVL around the circuit.

$$-1/sc\ I\,(s) - I\,(s)\,R + Vi\,(s) = 0$$

$$Is = \frac{Vi(s)}{\left[R + \dfrac{1}{sc}\right]}$$

$$V_o(s) = I(s)R = \frac{Vi(s)}{\left[R + \dfrac{1}{sc}\right]} \times R$$

$$V_o(s) = \frac{Vi(s)R}{\dfrac{scR+1}{sc}} \Rightarrow \frac{Vi(s)R}{1} \times \frac{sc}{(scR+1)} = \frac{scRVi(s)}{scR+1}$$

$$A = \frac{V_o(s)}{V_i(s)} = \frac{1}{1 + \dfrac{1}{scR}} \Rightarrow \text{Transfer function}$$

Numerator and De-numerator divided by SCR applying sinusoidal input varying its frequency 0 to α, $S = jw$

$$A = \frac{V_o(j\omega)}{V_i(j\omega)} = \frac{1}{1 + \frac{1}{j\omega RC}j}$$

$$\frac{1}{j} = -j, \quad j^2 = -1$$

$$\omega = 2\Pi f$$

$$A = \frac{V_o(j\omega)}{V_i(j\omega)} = \frac{1}{1 + j\frac{1}{2\Pi fRC}} \quad \text{Frequency domain transfer function}$$

$$|A| = \left|\frac{V_o(j\omega)}{V_i(j\omega)}\right| = \frac{1}{\sqrt{1 + \left(\frac{1}{2\Pi fRC}\right)^2}} \qquad \theta = -\tan^{-1}\frac{1}{2\Pi fRC}$$

At lower cut-off frequency f_1,

$$|A| = \frac{1}{\sqrt{2}}$$

$$\frac{1}{\sqrt{2}} = \frac{1}{\sqrt{1 + \left(\frac{1}{2\Pi f_1 RC}\right)^2}}$$

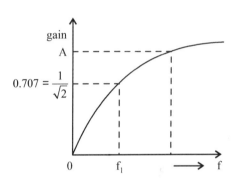

$$\frac{1}{2} = \frac{1}{1 + \left(\frac{1}{2\Pi f_1 RC}\right)^2}$$

$$2 = 1 + \left(\frac{1}{2\Pi f_1 RC}\right)^2$$

Fig 1.3 0 to f_1 – cut off Jone gain frequency plot

Equating the Denominators $2nfRC = 1$

$$f_1 = \frac{1}{2\Pi fRC} = \text{lower cut of frequency of high pass RC circuit}$$

$$A = \frac{V_o(j\omega)}{V_i(j\omega)} = \frac{1}{\sqrt{1 + \left(\frac{1}{2\Pi fRC}\right)^2}} = \frac{1}{\sqrt{1 + \left(\frac{f_1}{f}\right)^2}} \qquad \theta = \tan^{-1}\left(\frac{f_1}{f}\right)$$

1.4 STEP INPUT VOLTAGE

Let us consider that the step input voltage of Magnitude a voltage is applied as an input to the high pass RC circuit. When the input step is applied to the circuit, the current starts flowing instantaneously, then the capacitor changes exponentially and the current decays exponentially. Due to which the output voltage also decays exponentially. When capacitor charges equal to the input voltage level of voltage, current stops and the output voltage attains zero values in steady state conditions.

Let us mathematically analyse the output voltage as

$$V_o(t) = B_1 + B_2 e^{-t/\tau}$$

$B_1 \, B_2$, constants

τ is the time constant of the circuit

$$\tau = RC$$

The output voltage consists of two parts

1. B_1 is the steady state value of the output voltage

   $t \to \infty,$

   $$V_o(\infty) \to B_1$$

2. The transient part represented by expression decaying term $B_2 \, e^{-t/T}$

   The circuit is said to achieve steady state

   When the transient part completely dies out i.e., $t \to \infty$

   $$\text{Limt } t \xrightarrow{V_o(t)} \alpha(t) = t \xrightarrow{\lim} \alpha(B_1 + B_2 e^{-t/\tau})$$

   $$= B_1 \text{ as Lim } t \xrightarrow{\lim} e^{-t/\tau} = 0$$

   Let the steady state value of output voltage v_f

   $$B_1 = V_f$$

   To determine the B_2 (constant)

   $t = 0$ consider initial output voltage

   $t = 0$ be V_i

   $$V_o(t)\big|_{t=0} \Rightarrow B_1 + B_2 = V_i$$

   $$V_i = V_f + B_2$$

   $$B_2 = V_i - V_f$$

   Substituting the value B_1 and B_2

   $$V_o(t) = V_f + (V_i - V_f)e^{-t/\tau}$$

   Thus $t \to \infty$ the capacitor blocks d.c, hence the final steady state output voltage is zero

   $$V_f = 0$$

   The voltage across the capacitor cannot change instantaneously

   $t = 0^+$ i.e., just after $t = 0$

   The voltage across capacitor is zero. It can't change. Hence the output voltage at $= 0^+$ is same as the input voltage equal to A volt. When the capacitor is initially unchanged then the output is same as of input $t = 0^+$

$V_i = A$ voltage

$V_o(t) = V_f + (V_i - V_f) e^{-t/\tau}$

$\qquad = 0 + (A - 0)e^{-t/\tau}$

$\qquad = A\, e^{-t/\tau}$

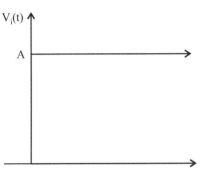

FIGURE 1.4 Step input

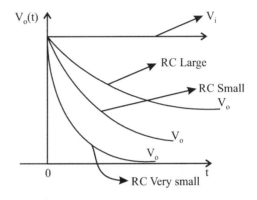

FIGURE 1.5 Step input for different time constants

.5 PULSE INPUT

An ideal pulse has the waveform shown in Figure (1.6). The pulse amplitude is V and pulse duration is t_p.

It has been mentioned earlier that the pulse is the sum of the two step voltages.

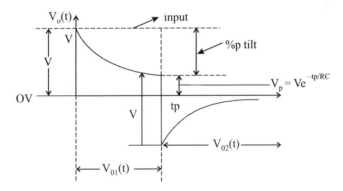

FIGURE 1.6 Pulse input waveform

So the response of the circuit $0 < t < t_p$ for the pulse input is same for a step input given by $V_{o1}(t) = Ve^{-t/RC}$.

At $t = t_p$ $\qquad V_{o1}(t) = Ve^{-t_p/RC} = V_p$

Now, consider the second part of the input for $t > t_p$. At $t = t_p$. As the input falls by V volts suddenly and the capacitor voltages can't change instantaneously, the output has to drop by a V volts to $V_p - V$

$$t = t_p \text{ i.e, } t_p^+$$

Hence the output drop by V from V_p at $t = t_p^t$ the capacitor voltage changes the output voltage decays exponentially to 0

For the second part of the pulse

$$t = t_p^t \quad V_{o2}(t_p^t) = V_p - V$$
$$V_{o2}(t_p^t) = Ve^{-tp/RC} - V$$
$$V_{o2}(t_p^t) = V(e^{-tp/RC} - V)$$

This is the initial output voltage for the second part of pulse

$$V_i = V(e^{-tp/RC} - 1)$$

The output voltage final value is zero

$$V_f = 0$$
$$V_{o2}(t) = V_f + (V_i - V_f) e^{-t/RC}$$
$$V_{o2}(t) = V(e^{-tp/RC} - 1) (e^{-(t-tp)/RC}$$

The output waveform RC $>> t_p$, RC comparable to t_p, and RC $<< t_p$ shown in figure 1.7 1.8, 1.9

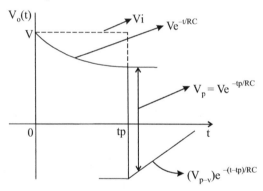

FIGURE 1.7 RC $>> t_p$

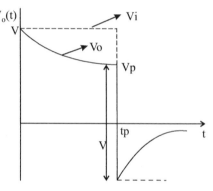

FIGURE 1.8 RC comparable to t_p

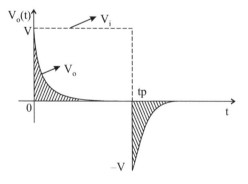

FIGURE 1.9 RC $<<t_p$

The response with large time constant RC ie, $RC/T_p >> 1$ is as shown in figure (1.7)

It can be observed that large time constant, the tilt is very small and undershoot also is very small, both the linear destruction are small. However the negative portion decreases very slowly

The response with small time constant $RC/t_p << 1$ is shown in Fig. (1.9). The output consists of a positive spike of amplitude V at the beginning of the pulse and a negative spike of the same size at the end of the pulse. This process of converting pulse into spikes using a circuit of small time constant is called peaking.

1.6 SQUARE-WAVE INPUT

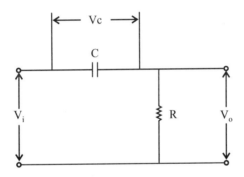

FIGURE 1.10 RC Circuit

Consider the various voltages present in high pass RC circuit as shown in the fig 1.10

q = charge on the capacitor

Apply Kirchhoff law $$c = \frac{q}{V} \quad V = \frac{q}{c}$$

$$V_i = V_c + V_o$$

$$V_i = \frac{q}{c} + vo$$

Differentiating the equation

$$\frac{dV_i}{dt} = \frac{1}{c}\frac{dq}{dt} + \frac{dv_o}{dt} \qquad\qquad i = \frac{dq}{dt}$$

$$\frac{dV_i}{dt} = \frac{1}{c}(i) + \frac{dV_o}{dt}$$

$$Vo = iR \quad i = \frac{V_o}{R}$$

Substituting in equation

$$\frac{dV_i}{dt} = \frac{V_o}{RC} + \frac{dV_o}{dt}$$

Both sides multiplied by the dt

$$dV_i = \frac{V_o}{RC} dt \, dV_o$$

Integrating the time period from 0 to T

$$\int_0^T dV_i = \frac{1}{RC} \int_0^T V_o \, dt + \int_0^T dV_o$$

$$[V_i]_0^T = \frac{1}{RC} \int_0^T V_o \, dt + [V_o]_0^T$$

$$V_i(T) - V_i(0) = \frac{1}{RC} \int_0^T V_o \, dt + V_o(T) - V_o(0)$$

Under steady-state conditions, the output waveform is repetitive with a time period T

$$V_i(T) = V_i(0) \text{ and } V_o(T) = V_o(0)$$

Hence $\int_0^T v_o(t)dt = 0$. This integral represents this area under the output waveform over one cycle i.e, the average value of output response, substituting the equations.

$$\frac{1}{RC} \int_0^T V_o \, dt = 0$$

The average level of the steady state output signal is always zero

[1] The average level of the output signal is always zero irrespective of the average level of the input. The output must extend in both positive and negative direction with respect to the zero voltage axis and area of the part of the waveform above the zero axis must equal the area below the zero axis.

[2] When input changes continuously by amount V, the output also changes by the same amount in the same direction.

[3] During any finite time interval where the input maintains a constant level, the output decays exponentially towards zero voltage.

They are in the limiting case, when the ratios RC/T_1 and RC/T_2 are both very large with respect to unity, the output waveform is exactly same as the input.

Now, consider the extreme case when RC/T_1 and RC/T_2 are very small as compared to unity.

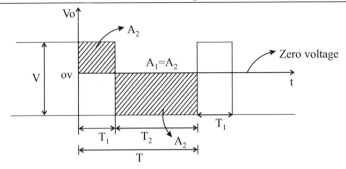

FIGURE 1.11

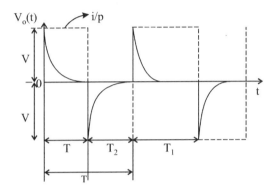

FIGURE 1.12 The high pass RC circuit with small time constant producer spikes circuit

Under steady state condition the capacitor charger and discharges to the same voltage level in each cycle.

For $0 < t < T_1$ the output is given by $V_{o1} = V_1 e^{-t/RC}$

$$\text{At } t = T_1 \quad Vo_1 = V_1^1 = V_1 e^{-T_1/RC}$$

For $T_1 < t < T_1 + T_2$ the output is $V_{o2} = V_2 e^{-(t-T_1-)/RC}$

$$\text{At } t = T_1 + T_2 \,, Vo_2 = V_2^1 = V_2 e^{-T_2/RC}$$

$$V_1^1 - V_2 = V \text{ and } V_1 - V_2^1 = V$$

Expression for the percentage tilt:

The Tilt is defined as the decay in the amplitude of the output voltage wave when the input maintains its level constant.

Mathematically the percentage tilt p is defined as

$$p = \frac{V_1 - V_1^1}{\text{input amplitude}} \times 100$$

When the time constant RC of the constant is very large compared to the period of the input waveform RC>>T

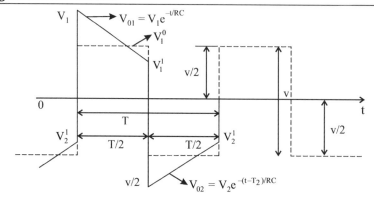

FIGURE 1.13 Tip tilt of a symmetrical square wave when RC >>T

For a symmetrical square wave with zero average value

$$V_1 = -V_2, \text{ i.e, } V_1 = |V_2|, \ V_1^1 = -V_2^1 \text{ i.e, } V_1^1 = |V_2|$$

and $T_1 = T_2 = T/2$

RC >>T shown in figure 1.13

$$V_1^1 = V_1 e^{-T/2RC} \text{ and } V_2^1 = V_2 e^{-T/2RC}$$

$$V_1 - V_2^1 = V$$

$$V_1 + V_1^1 = V \hspace{3cm} [\text{Note } - V_2^1 = V_1^1]$$

$$V_1 + V_1 e^{-T/2RC} = V$$

$$V_1 = \frac{V}{1 + e^{-T/2RC}} \text{ or (a) } V = V_1(1 + e^{-T/2RC})$$

$$\% \text{ tilt } p = \frac{V_1 - V_1^1}{V/2} \times 100\%$$

Input amplitude = v/2

$$= \frac{V_1 - V_1 e^{-T/2RC}}{V(1 + e^{-T/2RC})} \times 200\%$$

$$= \frac{1 - e^{-T/2RC}}{1 + e^{-T/2RC}} \times 200\%$$

When the time constant is very large T/2RC <<1

$$p = \frac{1 - \left[1 + (-T/2RC) + (-T/2RC)^2 \dfrac{1}{2!} \right]}{1 + 1 + (-T/2RC) + (-T/2RC)^2 \dfrac{1}{2!}} \times 200\%$$

$$\% \, p = \frac{1 - (1 - T/2RC)}{1 + (1 - T/2RC)} \times 200\% \qquad\qquad e^{-T/2RC} = 1 - \frac{T}{2RC}$$

$$p = \frac{\dfrac{T}{2RC}}{2} \times 200\%$$

$$= \frac{T}{2RC} \times 100\% = \frac{\pi f_1}{f} \times 100\%$$

$f_1 = \dfrac{1}{2\pi RC}$ is the lower cut off frequency of the high pulse RC circuit

1.7 RAMP INPUT

A waveform which is zero for $t < 0$ and which increases linearly with the time for $t > 0$ is called ramp (or) sweep voltage. Ramp input can be mathematically written as

$$Vi(t) = \begin{cases} 0 \text{ for } t < 0 \\ \alpha t \text{ to } t > 0 \end{cases}$$

Where α is the slope of the ramp

$$V_i = \frac{q}{c} + V_0 \Rightarrow V_i(t) = \frac{q}{c} + V_0(t)$$

$$Vi(t) = \alpha t = \text{input ramp}$$

$$\alpha t = \frac{q}{c} + V_0(t)$$

Differentiating the equation both side w.r.t t

$$\alpha = \frac{dq}{cdt} + \frac{dV_o}{dt}$$

$$\left[\text{Note } \frac{dq}{dt} = i, \frac{dq}{dt} = \frac{V_o}{R}, \, V_o = iR, \, i = \frac{V_o}{R} \right]$$

Substituting in the equation

$$\alpha = \frac{V_o}{RC} + \frac{dV_o}{dt}$$

Initially capacitor is zero $V_o(0) = 0$

take Laplace form

$$\frac{dV_o(t)}{dt} + \frac{V_o(t)}{RC} = \alpha$$

$$V_o(s)\, S + \frac{1}{RC} V_o(s) = \frac{\alpha}{S}$$

$$\left[S + \frac{1}{RC} \right] V_o(s) = \frac{\alpha}{S}$$

$$V_o(s) = \frac{\alpha}{S\left(S + \dfrac{1}{RC} \right)}$$

$$V_o(s) = \alpha RC \left[1/S - \frac{1}{S + 1/RC} \right]$$

$$V_o(t) = \alpha RC \left[1 - e^{-t/RC} \right]$$

$$V_o(t) = 0 \quad t = 0$$

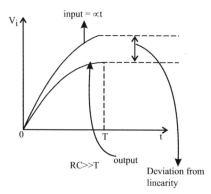

FIGURE 1.14 Deviation from linearity

$$V_o(t) = \alpha RC \left[1 - \left\{ 1 + (-t/RC) + \frac{(-t)^2}{(RC)^2 2!} + \frac{(-t)^3}{(RC)^3 3!} + \right\} \right]$$

$$= \alpha RC \left[t/2RC - \frac{t^2}{2(RC)^2} \right] = \frac{\alpha t - \alpha t^2}{2RC} = \alpha t \left[t - \frac{t}{2RC} \right]$$

The falling away of output from input is called deviation from linearity

This departure of output from linearity is called the trangenmussion error denoted as e_t.

$$e_t = \left. \frac{Vi - Vo}{Vi} \right|_{t=T} = \frac{\alpha t - \alpha t \left(1 - \dfrac{t}{2RC} \right)}{\alpha t} = \frac{T}{2RC} = \pi f_1 T$$

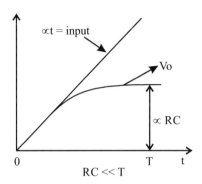

FIGURE 1.15 RC << T

.8 EXPONENTIAL INPUT (HPF)

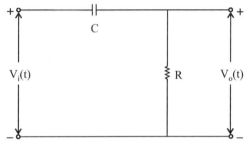

FIGURE 1.16 HPF

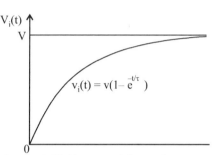

FIGURE 1.17 Exponential waveform

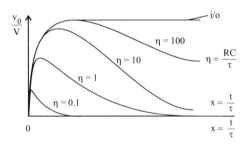

FIGURE 1.18 Output waveform

Let us consider the RC-high pass circuit and the exponential input denoted in the figure (1.16, 1.17, 1.18)

The exponential input can be expressed as

$$Vi(t) = V(1 - e^{-t/\tau})$$

Repeating similar steps of the previous sections

$$\frac{dV_i(t)}{dt} = \frac{V_o(t)}{RC} + \frac{dV_o(t)}{dt}$$

Substituting the equation for exponential input w   consider 'w'

$$\frac{V}{\tau} e^{-t/T} = \frac{V_o(t)}{RC} + \frac{dV_o(t)}{dt}$$

The intial output is zero as the intial voltage on the capacitor is zero.

$V_o(0) = 0$ which makes the Laplace transform approach suitable to solve the above differential equations the equation is rewritten as

$$\frac{dv_o(t)}{dt} + \frac{v_o(t)}{RC} = \frac{V}{\tau} e^{-t/\tau}$$

Taking Laplace transform both sides

$$\left[S + \frac{1}{RC}\right] v_o(s) = \frac{V}{\tau} \frac{1}{\left(s + \frac{1}{\tau}\right)}$$

$$v_o(s) = \frac{V}{\tau} \frac{1}{\left(s + \frac{1}{\tau}\right)\left(S + \frac{1}{RC}\right)}$$

$$V_o(s) = \frac{V}{\tau} \frac{1}{\left(\frac{1}{\tau}\right) - \left(\frac{1}{RC}\right)} \left[\frac{1}{S + RC} - \frac{1}{S + \frac{1}{\tau}}\right]$$

$$v_o(t) = \frac{VRC\tau}{\tau(RC - \tau)} \left(e^{-t/RC} - e^{-t/\tau}\right)$$

$$v_o(t) = \frac{VRC}{RC - \tau} \left(e^{-t/RC} - e^{-t/\tau}\right)$$

$$v_o(t) = \frac{\frac{VRC}{\tau}}{\frac{RC}{\tau} - 1} \left(e^{-t/RC} - e^{-t/\tau}\right)$$

x and n defined as

$$x = \frac{t}{\tau} \qquad\qquad n = \frac{RC}{\tau}$$

Do note that RC is the circuit time constant and τ is the input time constant

X may be Normalised time and n interpreted as the Normalised time constant

$$\frac{x}{n} = \frac{t}{RC}$$

The modified expression can be written as

$$v_o(t) = \frac{V_n}{n-1}\left(e^{-x/n} - e^{-x}\right), \qquad n=1$$

$$v_o(t) = \frac{V_n}{n-1}\left(e^{-x/n} - e^{-x}\right), \qquad n\neq1$$

it make use of L hospital rule

$$v_o(t) = \frac{\lim\limits_{n \to 1}\dfrac{d}{dn}\left[V_n(e^{-x/n} - e^{-x})\right]}{\lim\limits_{n \to 1}\dfrac{d}{dn}(n-1)}$$

$$v_o(t) = \lim\limits_{n \to 1}\frac{\left[V(e^{-x/n} - e^{-x}) + V_n(-e^{-x/n})\left[\dfrac{-x}{n^2}\right]\right]}{1}$$

$$v_o(t) = \lim\limits_{n \to 1}\left[V(e^{-x/n} - e^{-x}) + V_n(-e^{-x/n})\left[\dfrac{-x}{n^2}\right]\right]$$

$$v_o(t) = Vxe^{-x}$$

It conclude our derivation by starting that the response of the RC high pass circuit for as exponential waveform is given by

$$v_o(t) = \frac{Vn}{n-1}(e^{-x/n} - e^{-x}) \qquad \text{for } n\neq1$$

$$v_o(t) = Vxe^{-x} \qquad \text{for } n=1$$

[1] When n is large the response has larger peak amplitude as well as a wider pulse width.

[2] Similarly when the n response is smaller and has as smaller peak amplitude provided the width of the pulse is narrow n has an effect on both peak value and the width of the output pulse.

1.9 SINUSOIDAL INPUT

The analysis of the High pass RC circuit to sinusoidal input is obtained using Laplace transform approach applying KVL to the circuit.

$$-I(s)\frac{1}{sc} - I(s)R + Vi(s) = 0$$

$$Vi(s) = \frac{1(s)}{SC} + I(s)R$$

$$Vi(s) = I(s)\left[\frac{1}{SC} + R\right]$$

$$I(s) = \frac{Vi(s)}{SC + R}$$

$$Vo(s) = I(s)R = \frac{Vi(s)}{SC + R} \times R$$

$$\frac{Vo(s)}{Vi(s)} = \frac{R}{R + \dfrac{1}{SC}} = \frac{1}{1 + \dfrac{1}{S + RC}} = \text{Tranfer function}$$

Frequency varies from 0 to ∞ s replaced by jω

$$\frac{Vo(j\omega)}{Vi(j\omega)} = \frac{1}{1 + \dfrac{1}{j\omega RC}}$$

$$\frac{1}{j} = -1 \qquad\qquad \omega = 2\pi f$$

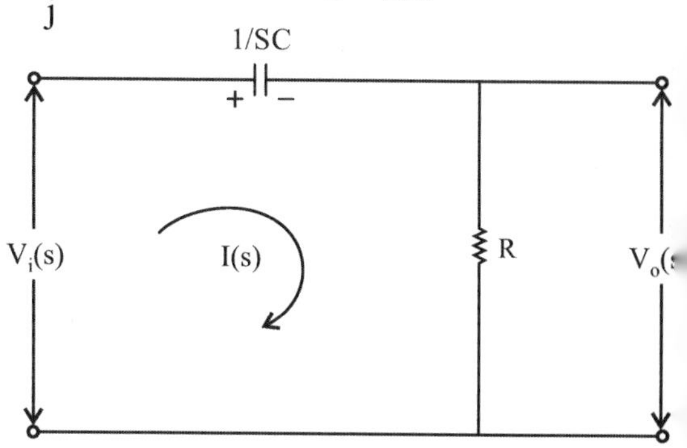

$$\frac{Vo(j\omega)}{Vi(j\omega)} = \frac{1}{1 - \dfrac{1}{j2\pi fRC}}$$

$$\frac{1}{1 + \dfrac{j}{2\pi fRC}} \Rightarrow$$

Frequency domain transfer function

$$A = \left|\frac{Vo(j\omega)}{Vi(j\omega)}\right| = \frac{1}{\sqrt{1 + \left(\dfrac{1}{2\pi fRC}\right)^2}}$$

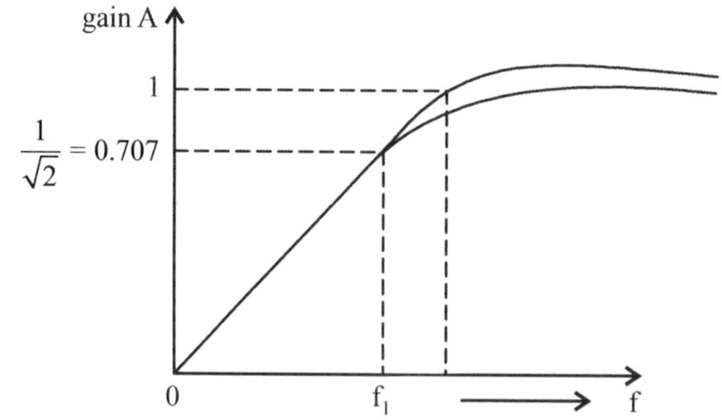

FIGURE 1.19 High pulse RC circuit

FIGURE 1.20 output waveform

Frequency increases the gain A approaches to unity. Initially output increases as the frequency increases and becomes equal to input at high frequency. As $f \to \infty, A \to 1$. To allow high-frequencies to pass.

A gain is $1/\sqrt{2}$ is called lower cout of frequency f_1 of the circuit.

$0 - f_1$ is cut off/zone

$$\frac{1}{\sqrt{2}} = \frac{1}{\sqrt{1 + \left(\frac{1}{2\pi f_1 RC}\right)^2}}$$

$$\frac{1}{2} = \frac{1}{1 + \left(\frac{1}{2\pi f_1 RC}\right)^2}$$

$$2 = 1 + \left(\frac{1}{2\pi f_1 RC}\right)^2$$

$$2\pi f_1 RC = 1 \qquad\qquad f_1 = \frac{1}{2\pi RC} = \text{lower cut off frequency}$$

.10 HIGH PASS RC CIRCUIT AS A DIFFERENTIATOR

For high pass RC circuit of time constant is very small in comparison with the time required for the input signal to make an appreciable change, the circuit is called differentiator.

Under this case, the drop across R is negligible compared to drop across C. Hence the total input vi(t) appears across C.

The current i is given

$$i(t) = C\frac{dvi(t)}{dt}$$

Hence the output which drops across R is

$$V_0 = iR$$

$$V_0(t) = RC\frac{dvi(t)}{dt}$$

The output is proportional to the derivative of the input. A criteria for good differentiation in terms of steady sate sinusoidal analysis is that if a sinusoidal is applied to the high pass RC circuit, the output will be a sine wave shifted by a leading angle θ such that $\tan\theta = \frac{\alpha c}{R}\frac{1}{wRC}$ the output will be proportional to sin $(w_t + \theta)$. In order to have true differentiation we must obtain cos w_t. In other words θ must be equal to 90^0. This result can be obtained only if R = 0 or C = 0. However if ω RC = 0.01, then $1/\omega CR$ = 100 and θ = 89.4^0 and for some applications this may be close enough to 90^0.

If the peak value of input is V_m, the output is

$$V_o = \frac{V_m R}{\sqrt{R^2 + \frac{1}{w^2 C^2}}} \sin(\omega t + \theta)$$

and if $\omega RC \ll 1$, then the output is approximately $V_m \omega RC \cos\omega t$. This results agrees with the expected value RC $\dfrac{dvi(t)}{dt}$. If ωRC = 0.01 then the output amplitude is 0.0 times the input amplitude.

These facts prove that with a small time constant the high pass RC circuit behaves as a differentiator.

The time constant RC of the circuit should be much smaller than the time period of the input signal RC<<T.

Application: RC>>T is employed in R-C completely of amplifier where distortion and differentiation of waveform is to be avoided, multi libratory, flip flap

1.11 LOW-PASS RC CIRCUIT

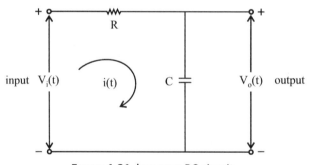

FIGURE 1.21 low pass RC circuit

Fig.1.21 shows a low pass RC circuit. The circuit passes the low frequencies readily, but attenuates high-frequencies because the reactance of the capacitor C decreases with increasing frequency. At very high frequencies the capacitor acts as virtual short-circuited and the output fall to zero. Thus, the high frequencies get attenuated. At zero frequency the reactance of the capacitor is infinity (capacitor is open circuit).

Sinusoidal input:

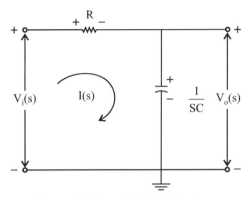

FIGURE 1.22 low RC circuit Laplace

If the input voltage is sinusoidal Vi(t) expressed as,

$$Vi(t) = V_m \sin \omega t$$

It can make use of the Laplace transform and analyse the circuit in s-domain. Since there is no change on the capacitor.

Applying KVL to the circuit as shown in figure

We can write

$$Vi(s) - I(s)R = \frac{1(s)}{SC} = 0$$

$$Vi(s) = I(s)R + \frac{1(s)}{SC}$$

$$Vi(s) = I(s)\left[R + \frac{1}{SC}\right]$$

$$I(s) = \frac{Vi(s)}{\left[R + \dfrac{1}{SC}\right]}$$

$$Vo(s) = I(s) \times \frac{1}{SC}$$

I(s) is substituting the Vo(s)

$$\frac{Vo(s)}{Vi(s)} = \frac{1}{\left[R + \dfrac{1}{SC}\right]} \times \frac{1}{(SC)} = \frac{1}{(SCR + \dfrac{\cancel{SC}}{\cancel{SC}})}$$

$$\frac{V_o(s)}{V_i(s)} = \frac{1}{1 + SRC} \quad \text{Transfer function}$$

For analysing frequency response replace S by $j\omega$

$$\frac{V_o(j\omega)}{V_i(j\omega)} = \frac{1}{1+j\omega RC} = \frac{1}{1+j2\pi fRC}$$ Frequency domain of transfer function

$$A = \left|\frac{V_o(j\omega)}{V_i(j\omega)}\right| = \frac{1}{\sqrt{1+(2\pi fRC)^2}} = \text{gain of the circuit}$$

At the upper is off frequency f_2, $|A| = \dfrac{1}{\sqrt{2}}$

$$\frac{1}{\sqrt{2}} = \frac{1}{\sqrt{1+(2\pi fRC)^2}}$$

$$\frac{1}{2} = \frac{1}{1+(2\pi f_2RC)^2} \quad \text{Equating denominator}$$

$$2 = 1+(2\pi f_2RC)^2$$

$$f_2 = \frac{1}{2\pi RC} = \text{upper cut off frequency}$$

cut off zone and from f_2 on wards

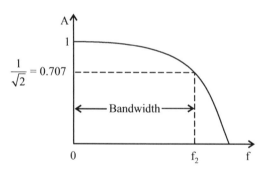

FIGURE 1.23 output wave form

The magnitude of the steady state gain A and the angle θ by which output leads the input is given by

$$A = \frac{1}{1+j\left(\dfrac{f}{f_2}\right)} \quad \text{and} \quad |A| = \frac{1}{1+\left(\dfrac{f}{f_2}\right)^2}$$

$$\theta = -\tan^{-1}\left(\frac{f}{f_2}\right)^2 \qquad\qquad f_2 = \frac{1}{2\pi RC}$$

It can explain output signal i $V_o(t) = AV_m\sin(\omega t + \theta)$, hence the phase angle θ is Negative

1.12   STEP VOLTAGE INPUT

Consider the step input voltage of magnitude A is applied to the low pass RC circuit having a time constant RC. A step voltage V (t) can be mathematically written as

$$V(t) = \begin{cases} 0 & \text{for } t > 0 \\ V & \text{for } t \geq 0 \end{cases}$$

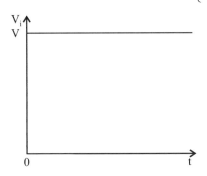

FIGURE 1.24 Step input

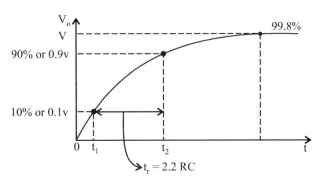

FIGURE 1.25 Step response Low pass RC current

If the capacitor is initially uncharged when a step input voltage is applied. The voltage across the capacitor can't charge instantaneously the output will be zero at t = 0. When the capacitor charges the output, voltage rises exponentially towards the steady state value V with.

Let V^1 is the initial voltage across the capacitor

Writing KVL around loop

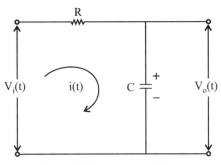

FIGURE 1.26 Low pass RC circuit

$$Vi\ (t) = i(t)\ R + \frac{1}{c}\int i(t)dt$$

Differentiating the equation

$$\frac{dVi(t)}{dt} = \frac{Rdi(t)}{dt} + \frac{1}{c}i(t)$$

Note

$$Vi(t) = V, \quad \frac{dVi(t)}{dt} = 0$$

$$\frac{di}{dt} = (\text{final value-intional value})$$

$$L(t) = 1/s$$

Take Laplace transform both side

$$L(1/t) = s$$

$$0 = \frac{Rdi(t)}{dt} + \frac{1}{C}i(t)$$

$$R[SI(s) - I(0^+)] + \frac{1}{c}I(s)$$

$$L(I(t)) = I(s)$$

$$I_{0^+} = Is\left[s + \frac{1}{RC}\right]$$

$$L\left[\frac{di(t)}{dt}\right] = SI(s) - (I_0^+)s$$

Final   Initial

The initial current I_0^+ is given by

$$I_0^+ = \frac{V - V^1}{R}$$

$$I(s) = \frac{I(0^+)}{S + \dfrac{1}{RC}} = \frac{V - V^1}{R(S + \dfrac{1}{RC})}$$

$$V_0(s) = Vi(s) - I(s)R$$

$$= V/S - \frac{(V - V^1)}{R(S + \dfrac{1}{RC})}R = \frac{V}{S} - \frac{V - V^1}{S + \dfrac{1}{RC}}$$

Taking Inverse Laplace transform both sides

$$V_0(t) = V - (V - V^1)\,e^{-t/RC}$$

V is the final voltage (v final) when the capacitor is charged

V^1 is the internal voltage across the capacitor

$$V_0(t) = V\ \text{final} - (V\ \text{final} - V\ \text{final})e^{-t/RC}$$

The capacitor fanatically uncharged than

$$V_0(t) = V(1 - e^{-t/RC})$$

Expression for rise time:

The rise time t_r is defined as the time it take the voltage to rise from 0.1 to 0.9 of it final value. It gives an indication of how fast the circuit can respond to a discontinuit in voltage.

Assuming the capacitor is initially uncharged.

The time required for the output to achieve 10% of its final value can be obtained

$$Vo\ (t) = V\ (1 - e^{-t/RC})$$

at $\qquad$ $t = t_1$ $\qquad$ $Vo(t) = 10\%\ (or)\ V = 0.1\ V$

$$0.1\ V = V\ (1 - e^{-t/RC})$$

$$0.1 = 1 - e^{-t/RC}$$

$$e^{-t1/RC} = \ln 0.9$$

$$\frac{-t_1}{RC} = |n\,(0.9) \qquad\qquad\qquad t_1 = 0.1\ RC$$

Similarly the time required for the o/p to achieve 90% of its final value output

$$t = t_2 \qquad Vo\ (t)\ ⊔\ 90\%\ (or)\ V = 0.9V$$

$$0.9\ \not{V} = \not{V}\ (1 - e^{-t_2/RC})$$

$$0.9 = 1 - e^{-t_2/RC}$$

$$e^{-t_2/RC} = 0.1$$

$$\frac{-t_2}{RC} = |n\,(0.1)$$

$$t_2 = 2.3\ RC$$

$$t_r = t_2 - t_1$$

rise time $t_r = 2.3\ RC - 0.1\ RC = 2.2\ RC$

Relation between upper 3 dB frequency and rise time

$$f_2 = \frac{1}{2\pi RC}\ (or)\ RC = \frac{1}{2\pi f_2}$$

$$\text{Rise time} = 2.2\ RC = \frac{2.2}{2\pi f_2} = \frac{0.35}{f_2} = \frac{035}{BW}$$

Rise time is inversely proportional to the upper 3 dB frequency and directly proportional to the time constant RC.

$$\tau = \text{time constant} = RC \text{ in RC circuits}$$

1.13 PULSE INPUT VOLTAGE

Consider the pulse input voltage having pulse width t_p, applied as input to the RC circuit the pulse sum the two step voltages the response to a pulse for times less than the pulse width t_p is the same as that for a step input because pulse signal is same as the step input for $t < t_p$. However at the end of the pulse as the input become zero. The output also drops exponentially to zero as capacitor voltage falls exponentially to zero as the input becomes zero.

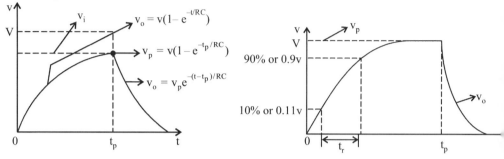

FIGURE 1.27 $RC \gg t_p$ FIGURE 1.28 $RC < t_p$

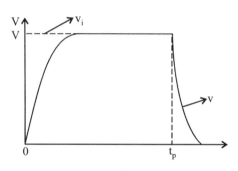

FIGURE 1.29 $RC \ll t_p$

Output for pulse input is given by

$$V_{out} = V(1 - e^{-t/RC}) \quad t < t_p$$

$$V_{out} = V(1 - e^{-t/RC}) = V_p \text{ (say)}$$

$t = t_p$, input voltage becomes zero but the voltage across a capacitor can't charge instantaneously. Output remain the same as it is $t = t_p$. After that capacitor starts getting discharged through resistance R and voltage across it drops exponentially to zero.

The output voltage $t > t_p$

$$V_{out} = V_p \, e^{-(t-tp)/RC}$$

The above equation, discharging equation of capacitor delayed by time t_p. The output voltage must be decreasing towards to zero.

The output voltage will always extend beyond pulse width t_p. This is because charge stored on capacitor during pulse cannot leak off instantaneously.

To minimize the distortion, the resistance must be small compared with the pulse width t_p.

$$f_2 = \frac{1}{t_p} \quad t_r = 0.35f_2$$

The upper 3-dB frequency f_2 is chosen equal to the reciprocal of the pulse width t_p.

.14 SQUARE WAVE INPUT

Consider a periodic waveform whose instantaneous value is constant at 'V' with respect to ground for a V"T_1 and changes abruptly for time T_2 at regular interval $T = T_1 + T_2$. A reasonable reproduction of the input is obtained if the resistance tr is small compared with the pulse width.

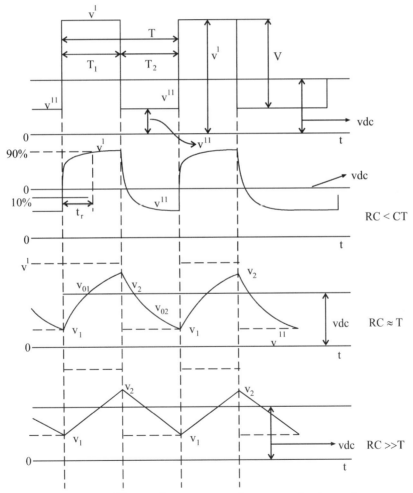

FIGURE 1.30 Different time constant of square output waveform

The steady state response is drawn in fig (b)

If the time constant RC is comparable with the period of the input square wave, the output will have the appearance shown in fig (c) if the time constant is very large compared with the input wave period, the output is exponentially linear as illustrated in fig (d).

In fig (c) rising portion of the equation

$$V_{o1}(t)V' - (V' - V_2)e^{-t/RC}$$

Where V_2 is the voltage across the capacitor at $t = 0$ and V' is the level of the capacitor charge.

So, the output voltage for $0 < t < T_1$

$$V_{o1}(t)V' + (V_1 - V')e^{-t-/RC}$$

Similarly for $T_1 \leq t \leq T_2$, if intial voltage across capacitor is V_2 and input voltage i constant at it V'' and output voltage $= V_{o2}$

$$V_{o2} - V'' + (V_2 - V'')e^{-(t-T_1)RC}$$

$$t = T_1, \quad V_{o1} = V_2$$

$$t = T_1 + T_2, \quad V_{o1} = V_2 = V^1 + (V_1 - V')e^{-T_1/RC}$$

then $\quad t = T_1 \quad V_{o2} = V_1$

$$V_2 = V' + (1 - 0^{-T_1/RC}) + V_1 e^{-T/RC}$$

$$V_1 = V'' + (V_2 - V'')e^{-T_2/RC}$$

For symmetrical wall

$$V_1 = -V_2, \quad V' = -V''$$

$$T_1 = T_2 = T/2$$

$$V_1 = -V' + (-V_1 + V')e^{-T/RC_2}$$

$$V_1 = V' - V_1 e^{-T/RC_2} + V'e^{-T/RC_2}$$

$$V_1 = (1 + e^{-T/2RC}) = V'(e^{-T/RC_2} - 1)$$

$$V_1 = \frac{V'(e^{-T/RC_2} - 1)}{e^{-T/RC_2} + 1}$$

Input square wave of peak to peak voltage V

$$V' = V/2$$

$$\frac{V}{2} = \frac{(e^{-T/RC_2} - 1)}{(e^{-T/2RC} + 1)}$$

$$V_2 = \frac{-V(e^{-T/2RC} - 1)}{2(e^{-T/2RC} + 1)}$$

$$= \frac{V}{2}\left[\frac{1 - e^{-T/2RC}}{1 + e^{-T/2RC}}\right]$$

1.15 EXPONENTIAL INPUT : (LPF)

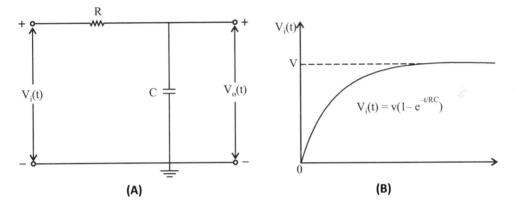

FIGURE 1.31 The exponential wave form (a,b)

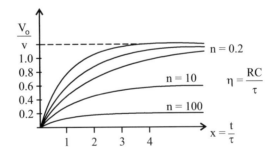

FIGURE 1.32 The output of the RC low pass circuit

apply KVL to the circuit we can write

$$Vi(t) = i(t)R + \frac{1}{c}\int_0^t i(t)dt$$

$$V_i(t) = RC + \frac{dV_o}{dt} + V_o(t)$$

$$V(1-e^{-t/\tau}) = RC \times \frac{dv_o(t)}{dt} + V_o(t)$$

Apply Laplace transform both sides

$$\frac{V}{s} - \frac{V}{s+1/\tau} = RCSV_o(s) + V_o(s)$$

$$V_o(s) = \frac{1}{RC\tau s(s+1/\tau)(s+1/RC)}$$

$$V_o(s) = \frac{V}{S} + \frac{V}{(\frac{RC}{\tau}-1)(S+\frac{1}{\tau})} - \frac{V}{(1-\frac{\tau}{RC})(S+\frac{1}{RC})}$$

$$X = \frac{t}{\tau} \text{ and } n = \frac{RC}{\tau}$$

Substituting expression, we obtain the output waveform in both the cases.

$$V_o(t) = \frac{V}{S} + \frac{V}{(n-1)(S+\frac{1}{\tau})} - \frac{V}{(1-\frac{1}{n})(S+\frac{1}{RC})}$$

Taking in inverse Laplace transform on both side

$$V_o(t) = V\left[1 + \frac{1}{(n-1)}e^{-t/\tau} - \frac{n}{(n-1)}e^{-t/RC}\right]$$

$$V_o(t) = V\left[1 + \frac{1}{(n-1)}e^{-x} - \frac{n}{(n-1)}e^{-x/n}\right] \qquad \text{for } n \neq 1$$

This equation is not valid when $n = 1$

We can find the expression for output $n = 1$ by using L'Hospital Rule

$$V_o(t) = \underset{n \to 1}{Lim} \frac{\frac{d}{dn}\left[V(n-1+e^{-x}-ne^{-x/n})\right]}{\underset{n \to 1}{Limt}\frac{d}{dn}(n-1)}$$

$$V_o(t) = \underset{n \to 1}{Lim} \frac{\left[V(1-e^{-x/n}) - V_n\left[e^{-x/n}\left(\frac{-x}{n^2}\right)\right]\right]}{1}$$

$$V_o(t) = \underset{n \to 1}{Lim}\left[V(1-e^{-x/n}) - V_n\left[e^{-x/n}\left(\frac{-x}{n^2}\right)\right]\right]$$

$$V_o(t) = V(1+(1+x)e^{-x}) \qquad \text{for } n = 1$$

$\tau =$ is the input time constant

RC is the circuit time constant

x may be tread normalised time x may be interpreted as the normalised time constant

$$\frac{x}{n} = \frac{t}{RC}$$

$$V_o(t) = \left[1 + \frac{1}{n-1}e^{-x} - \frac{n}{n-1}e^{-x/n} \right] \qquad \text{for } n \neq 1$$

$$V_o(t) = 1\left(1 + (1+x)e^{-x} \right) \qquad \text{for } n = 1$$

If the time constant of this response is τ then the rise time of this exponential waveform can be written as $t_r = 2.2\tau$.

.16 LOW PASS RC CIRCUIT AS AN INTEGRATOR

For low pass RC circuit, if the time constant is very large when compared to the time required by the input signal to make an appreciable change compared, the circuit acts as an integrator. The voltage drops across C will be very small in comparison to the drop across R and it may consider that the total input appears across R, then the current is Vi(t)/R and the output signal across C is

$$V_o(t) = \frac{1}{C}\int i(t)dt = \frac{1}{C}\int \frac{vi(t)}{R}dt = \frac{1}{RC}\int vi(t)dt$$

Hence the output is proportional to the integral of the input

$$Vi(t) = \alpha t, \text{ the result is } \alpha t^2 / 2RC$$

$$Vo(t) = \frac{\alpha t^2}{2RC}$$

As time increases, the drop across will not remain negligible compared with that across R and the output will not remain the integral of the input. The output will change from quadrate to a linear function of time.

Low pass RC circuit time constant is very large in compression with the time required for the input signal the circuit acts as a integrator.

Integrator is almost invariably preferred over differentiation in analogue computer application. These resource are given below.

(i)   An integrator is less sensitive to noise voltage than a differentiator because of its limited band with.

(ii) It is more convenient to introduce initial conditions in an integrator.

(iii) The differentiator overloads the amplifier if the input changes rapidly. This is not the case for an integrator.

(iv) The gain of an integrator decrease as the frequency. Hence easy to stabilise. The gain of the differentiator increase as the frequency, hence suffers from the problem of stability.

1.17 ATTENUATORS

It consider now the simple resistance attenuator which is used to reduce the amplitude of single waveform the single resistance combination of fig (1.33) would multiply the input signal by the ratio a $= \dfrac{R_2}{R_1 + R_2}$ independent of frequency.

The potential decoder consisting of two resistances R_1 and R_2, used as an attenuator.

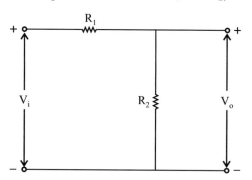

FIGURE 1.33 Simple attenuator

If the output of the attenuator the input capacitance C_2 of the amplifying will be the stray capacitor and attenuator, the resistor R_2 of the attenuator as shown in figure.1.31

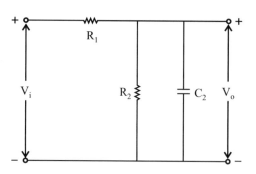

FIGURE 1.34 Actual attenuator

Attenuator equivalent circuit as shown in Fig 1.3

Thevenin voltage is A v_i which attenuated voltage and R is equal to the parallel combination of R_1 and R_2. Generally R_1 and R_2 are very large so that the nominal input impedance of the attenuator may be large enough to prevent loading down the input signal, the time constant RC_2 of the circuit is large which is totally unacceptable. Due to large time constant, resistance is also large which causes destruction. The high frequency components get attenuated. Hence attenuator no longer remains independent of the frequency.

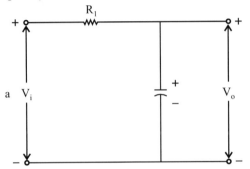

FIGURE 1.35 Attenuator equal at circuit

The attenuator may compensate so that, its attenuation is once again independent of frequency, by shunting R_1 by a capacitor C_1 as shown in figure 1.36.

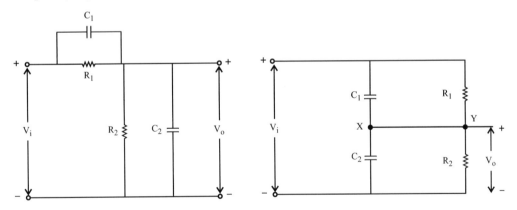

Figure 1.36 (a) (b) Compensated attenuator

The circuit can be redrawn such that the two resistors and two capacitors act as four across of a bridge figure (1.36 (b))

$$R_1C_1 = R_2C_2$$

Under the balanced condition no current can flow through the branched joining the terminal x and y. Hence, for calculating output, the branch x-y can be omitted under balanced bridge condition. This output is equal to v_i independent of frequency.

1.18 STEP INPUT RESPONSE

Let us find out the output waveform, when the step voltage is applied to the compensated attenuator. The step input has amplitude V, applied at t = 0 so the input change from 0 to V instantaneously at t = 0.

Now the voltage across C_1 and C_2 must change abruptly. But the voltage across capacitor cannot charge instantaneously if the current remain finite. Infinite current exists at t = 0. For an infinitesimal time so the finite charge $q = \int_{0-}^{0+} i(t)dt$ is delivered to each capacitor. So just after t = 0 ie at 0^+.

$$A = \frac{q}{c_1} + \frac{q}{c_2} \qquad\qquad C = \frac{Q}{V}$$

$$A = q\left[\frac{C_1 + C_2}{C_1 C_2}\right] \qquad\qquad V = \frac{Q}{C}$$

$$q = A / \left[\frac{C_1 + C_2}{C_1 C_2}\right]$$

Output voltage at $t = 0^+$ is voltage across the capacitor C_2 at $t = 0^+$

$$V_0(0^+) = \frac{q}{C_2}$$

$$V_0(0^+) = \frac{AC_1 \cancel{C_2}}{(C_1 + C_2)\cancel{C_2}}$$

$$V_0(0^+) = \frac{AC_1}{(C_1 + C_2)}$$

In the steady state as $t \to \alpha$ both the capacitors act as open circuited. Hence, the final value of the output voltage a totally by the resistor

$$V_0(\alpha) = \frac{AR_2}{R_1 + R_2}$$

For the perfect compensation

$$V_0(0^+) = V_0(\alpha)$$

$$\frac{C_2}{C_1 + C_2}\cancel{A} = \frac{R_2}{R_1 + R_2}\cancel{A}$$

$$(R_1 + R_2)\, C_1 = R_2\, (C_1 + C_2)$$

$$R_1 C_1 = R_2 C_2$$

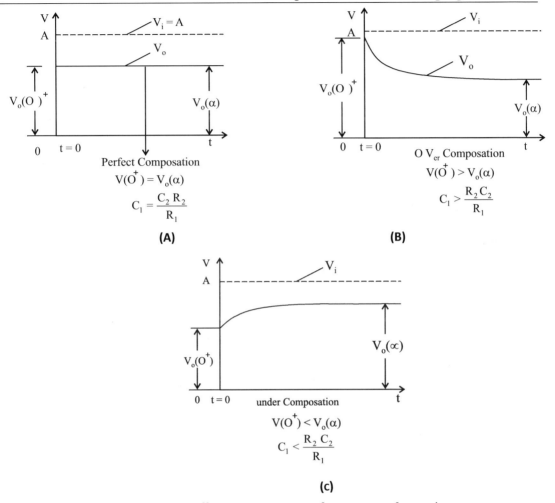

(A)

Perfect Composation
$$V(O^+) = V_o(\alpha)$$
$$C_1 = \frac{C_2 R_2}{R_1}$$

(B)

$O\ V_{er}$ Composation
$$V(O^+) > V_o(\alpha)$$
$$C_1 > \frac{R_2 C_2}{R_1}$$

(c)

under Composation
$$V(O^+) < V_o(\alpha)$$
$$C_1 < \frac{R_2 C_2}{R_1}$$

FIGURE 1.37 Different compensator of output waveform a, b, c

.19   HIGH PASS RC CIRCUIT

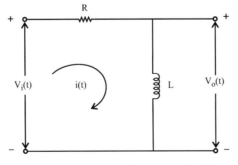

FIGURE 1.38 High pass RL circuit

By applying KVL we can write that

$$Vi(t) = VR(t) + VC(t)$$

$$Vi(t) = i(t) R + \frac{L di(t)}{dt}$$

Applying Laplace transform on both sides, we can write that

$$Vi(s) = RI(s) + SLI(s)$$

$$I(s) = \frac{Vi(s)}{R + LS}$$

$$V_o(t) + L\frac{di(t)}{dt} \qquad \text{we can write}$$

$$V_o(s) + LsI(s)$$

$$V_o(s) + Ls\frac{vi(s)}{R + LS}$$

$$\frac{V_o(s)}{Vi(s)} + \frac{LS}{R + LS}$$

$$G(s) = \frac{V_o(s)}{Vi(s)} + \frac{LS/R}{1 + LS/R}$$

G(s) as the transfer function of the circuit. Frequency function can be obtained G(f) by replacing $s = j\omega \; s = j2\pi f$

$$G(f) = \frac{1}{1 - j\dfrac{R}{2\pi fL}} = \frac{1}{1 - j(f_1/f)}$$

$$G(f) = \frac{1}{1 - (f_1/f)}$$

Hence f_1 represents the lower cut off frequency

$$f_1 = = \frac{1}{2\pi L/R} \qquad \text{G(f) in terms of } f_1$$

$$\text{we can write } A = \frac{1}{\sqrt{1 + (f_1/f)^2}} \qquad G(f) = |G(f)| \underline{|G(f)} = A\underline{|\phi}$$

$$\phi = \tan^{-1}(f_1/f)$$

Hence A is the magnitude; ϕ is the phase angle of the frequency function.

1.20 STEP INPUT VOLTAGE OF HIGH PASS RL CIRCUIT

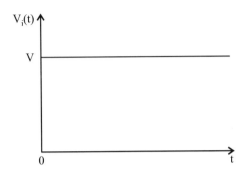

FIGURE 1.39 Step wave form

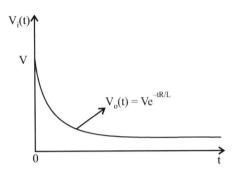

Fig 1.40 Out wave form RL HP circuit

Consider the RC high pass R_2 circuit as shown in fig 39, 40 applied to the step input to the high pass R_2 circuit the step function of amplitude RL can be mathematically writ written as Vi (t) = Vo(t) assume that the initial condition is over zero we know that the Laplace transform of the function

$$Vi(s) = \frac{V}{s}$$

The transfer function of the RC high pass circuit has been obtained

$$G(s) = \frac{Vo(s)}{Vi(A)} = \frac{Ls}{R + Ls} = \frac{S}{S + \dfrac{R}{L}}$$

$$Vo(s) = Vi(s)\,G(s) = \left(\frac{V}{S}\right)\frac{S}{S + \dfrac{R}{L}} = \frac{V}{S + \dfrac{R}{L}}$$

$$Vo(s) = \frac{V}{S + \dfrac{R}{L}}$$

In time domain equation can be written as

$$Vo(s) = Ve^{-Rt/L}$$

1.21 LOW PASS RL CIRCUIT

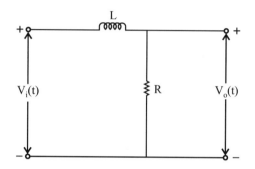

FIGURE 1.41 Low pass RL circuit

By applying KVL we can write that

$$Vi(t) = V_R(t) + V_L(t)$$

$$Vi(t) = i(t)R + \frac{Ldi(t)}{dt}$$

Applying Laplace transform on both side

$$Vi(s) = RI(s) + LSI(s) = Is(R+LS)$$

$$I(s) = \frac{Vi(s)}{R + Ls}$$

$$Vo(t) = i(t)\ R$$

We can write

$$Vo(s) = I(s)R$$

$$Vo(s) = R\frac{Vi(s)}{(R + Ls)}$$

$$G(s) = \frac{Vo(s)}{Vi(s)} = \frac{R}{R + Ls}$$

$$G(s) = \frac{1}{1 + \dfrac{LS}{R}} \Rightarrow \text{Transfer function of the circuit}$$

S is replaced by jw = j2 $\prod$ f

$$G(t) = \frac{1}{1 + j2\pi f(\dfrac{L}{R})}$$

$$G(t) = \frac{1}{1 + j(f / f_2)}$$

Hence f_2 is representing the upper cut off frequency

$$f_2 = \frac{1}{2\pi(L/R)}$$

The frequency function G(t) in terms of f_2

$$G(f) = |G(f)| \underline{|G(f)|} = A\underline{|\phi}$$

$$A = \frac{1}{\sqrt{1+(f/f_2)^2}}$$

$$\phi = \tan^{-1}(f/f_2)$$

Hence A is the magnitude

ϕ is the phase angle of the frequency function.

.22 STEP INPUT VOLTAGE OF LOW PASS RC CIRCUIT

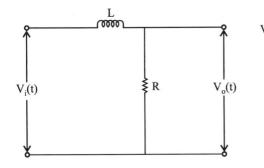

FIGURE 1.42 Low pass RC circuit

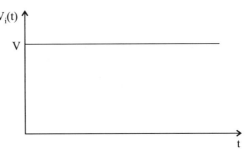

FIGURE 1.43 The step waveform of input

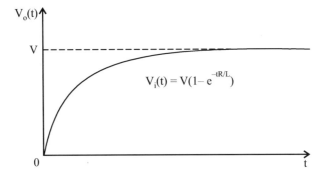

$$V_i(t) = V(1- e^{-tR/L})$$

FIGURE 1.44 Output wave form of step

Consider the low pass RC circuit indicated in fig 42, 43, 44, step input voltage i applied to the RL low pass circuit. The step function of amplitude V can be mathematically written as $Vi(t) = V_o(t)$.

Assume that the initial condition is zero. So that we can obtain the tranfer function o the circuit. We know the Laplace transform of the this function

$$Vi(s) = \frac{V}{s}$$

$$G(s) = \frac{Vo(s)}{Vi(s)} = \frac{R}{R+LS} = \frac{\dfrac{R}{L}}{S+\dfrac{R}{L}}$$

$$Vo(s) = Vi(s)\ G(s) = (\frac{V}{s}) \times (\frac{R/L}{S+R/L})$$

$$\left(\frac{R/L}{S(S+R/L)}\right) = V\left[\frac{1}{S} + \frac{1}{S+\dfrac{R}{L}}\right]$$

$$Vo(s) = V\left[\frac{1}{S} + \frac{1}{S+\dfrac{R}{L}}\right]$$

In time domain this equation can be written as

$$Vo(t) = V(1 - e^{-Rt/L})$$

1.23 RINGING CIRCUIT

The RLC circuit which produces nearly undamped oscillations is called ringing circuit. The RLC circuit undamped ratio ξ reduces, the oscillation response increases.

When ξ tends to zero the circuit oscillations for long time and performes many cycles, the oscillations reduces ϕ = is ringing circuit value, N is the number of cycles

$$Q = \pi N$$

$$N = \frac{Q}{\pi}$$

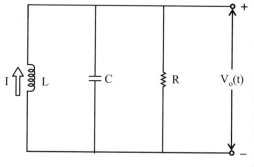

FIGURE 1.45 Ringing circuit

The ringing circuit as shown in figure 1.45

For initially capacitance is uncharged and inductor carries an initial current I.

When damping is made very small the output becomes undamped and takes the form of sine wave which on oscillated in magnetic energy gets stored in an inductor during one part of the cycle. It is converted into electrostatic energy stored in capacitor during next part of cycle.

Then amplitude of the oscillation is

$$\frac{1}{2}cv^2_{max} = \frac{1}{2}LI^2$$

$$V_{max} = 1\sqrt{\frac{L}{C}}$$

Application: The ringing circuit is used to generate the sequence of pulse.

1.24 RLC SERIES CIRCUIT

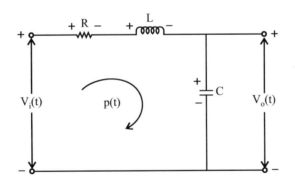

FIGURE 1.46 RLC series circuit

RLC series circuit is shown in fig 1.46

The output taken across capacitor 'C'

Applying KVL Loop

$$Vi(t) - i\,(t)\ R - \frac{Ldi(t)}{dt} - \frac{1}{c}\int i(t)dt = 0$$

Take Laplace transform of the above equation. Initially capacitor is unchanged, inductor current is zero

$$Vi(s) = I(s)R + LSI(s) + \frac{1}{c}\frac{I(s)}{S}$$

$$Vi(s) = I(s)(R + LS) + \frac{1}{cS}$$

$$I(s) = \frac{Vi(s)}{R + LS + \dfrac{1}{cS}}$$

From the circuit the output equation is

$$V_o(t) = \frac{1}{C}\int i(t)dt$$

$$V_o(s) = \frac{I(s)}{SC}$$

I(s) substitutes the above of output equation

$$V_o(s) = \frac{1}{SC}\left[\frac{VI(s)}{R + LS + \dfrac{1}{SC}}\right]$$

$$\frac{V_o(s)}{Vi(s)} = \frac{1}{S^2 LC + SRC + 1}$$

Numerator and denominator divided by the $\dfrac{1}{LC}$

$$\frac{V_o(s)}{Vi(s)} = \frac{\dfrac{1}{LC}}{S^2 + \dfrac{RS}{L} + \dfrac{1}{LC}}$$

The ratio of Vo(s) to Vi(s) is called transfer function in the circuit.

The equation obtained by equating denominator polynomial of a transform function is zero.

$$S^2 + \frac{R}{L}s + \frac{1}{LC} = 0$$

$$S_1, S_2 = \frac{-\dfrac{R}{L} \pm \sqrt{\left[\dfrac{R}{L}\right]^2 - \dfrac{4}{LC}}}{2} \qquad \text{Note: } ax^2 + bx + c \quad \frac{b \pm \sqrt{b^2 - 4ac}}{2a}$$

$$= -\frac{R}{2L} \pm \sqrt{\left[\frac{R^2}{2L}\right]^2 - \frac{1}{LC}}$$

Critical resistance RCr:

This resistance of value which reduce square root term to zero. Giving real, equal and negative roots.

$$\frac{RC}{2L} = \sqrt{\frac{1}{LC}}$$

$$RC_r = 2\sqrt{\frac{L}{C}}$$

(i) Damping ratio (ξ): it is denoted by Greek letter zeta (ξ)

The ratio of a dual resistance in the circuit to the critical resistant.

$$\xi = \frac{R}{RC_r} = \frac{R}{2}\sqrt{C/L}$$

(ii) Characteristic independence: he term $\sqrt{L/C}$ is calledd characteristic independence

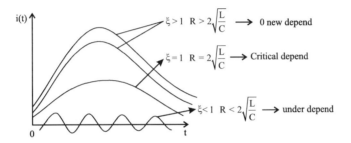

FIGURE 1.47 RLC Series circuit of current response

(iii) Natural frequency w(n)

$$\omega(n) = \frac{1}{\sqrt{LC}}$$

.25 RLC PARALLEL CIRCUIT

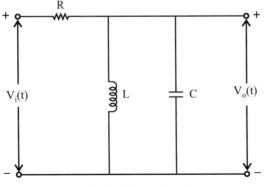

FIGURE 1.48 RLC parallel circuit

2

Non Linear Wave Shaping

2.1 INTRODUCTION

We studied in chapter 1 on linear wave shaping that non sinusoidal waveforms are transmitted through differentiator and integrator for altering their wave shapes. In this chapter some aspects of non linear wave shaping like clipping and clamping are taken up.

The clipper circuits are used to remove the unwanted portion of the waveform while clamper circuits are used to add (or) restore dc level to a signal. The clippers are also called slicers (or) voltage limiters or amplitude selectors while the clampers are also called dc restorer dc reinserter.

Non linear wave shaping circuits may be classified as clipping circuits and clamping circuits may be single level clipper (or) two level clippers.

Clamping circuits may be negative clamper with final (or) and without reference (or positive clamper with and without reference.

2.2 CLIPPER CIRCUIT (OR) LIMITER

The basic principle of clipper circuits is to remove the certain portion of the waveform below and above the certain level as per the requirements. These are used to clip of unwanted waveform without distorting the remaining part of the waveform are called clipper circuits (or) clippers.

Series clipper: when the diode is connected in series with load, such circuit is called series clipper

Parallel clipper: When the diode is connected in a branch which is parallel to the load, it is called parallel clipper.

2.3 SERIES NEGATIVE CLIPPER CIRCUIT

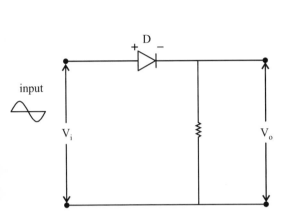

FIGURE 2.1 Negative series clipper

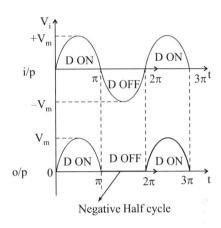

FIGURE 2.2 Negative half cycle is clipped off

Fig 2.1 shows a series Negative clipper circuit. For a positive half cycle the Diode D is forward biased. Hence the voltage wave form across R_L looks like a positive half cycle of the input voltage. While for a Negative half cycle Diode D is reversed biased and hence will not conduct at all. There will not be any voltage available across resistance R_L. Hence the Negative cycle of input voltage gets clipped off. The input and output wave is shown in fig.2.2.

Transfer charactistics of ideal diode

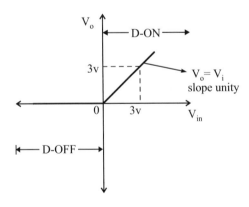

FIGURE 2.3 Transfer characteristics of ideal diode. The Mathematical equation for ideal diode is given by

$V_{in} \geq 0 \ V_o = V_{in}$ Diode is ON

$V_i < 0 \ \ V_o = 0$ Diode is OFF

Diode does not conduct till the input voltage is greater than the cut-in voltage (V_r) of the diode.

Silicon diode V_r is 0.7 V while $V_i \leq V_r$, the diode is OFF, $V_o = 0$ V

The mathematical equation for transfer character becomes

$V_i > 0.7$ V $\quad V_o = V_i - 0.7 \quad$ Diode is ON

$V_i \leq 0.7$ V $\quad V_o = 0 \quad\quad\quad$ Diode is OFF

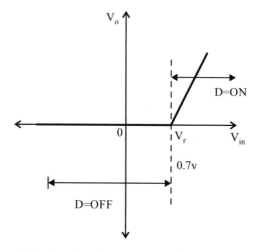

FIGURE 2.4 Transfer characteristics of practical diode

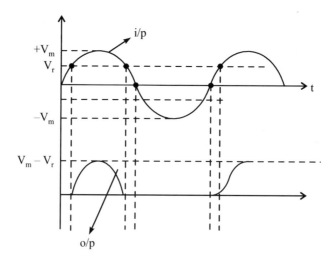

FIGURE 2.5 Input and out waveform

Diode is ON called transmitting region while the region for which diode is OFF is called clipping or limiting region

.4 SERIES POSITIVE CLIPPER CIRCUIT

Figure 2.6 shows positive series clipper circuit in which diode is direction opposite to that in negative series clipper which clips off positive part of the input can be obtained. It is called positive clipper.

Transfer characteristics:

Equation of the ideal diode transfer characteristics is

$V_i > 0$ $V_o = 0$ Diode is OFF

$V_i \le 0$ $V_o = V_{in}$ Diode is ON

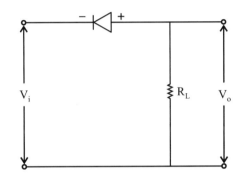

FIGURE 2.6 Transfer characteristics of ideal diode

Hence mathematical equation of practical diode with V_γ cut in voltage

$V_i \ge -0.7 \text{ V}$ $V_o = 0$ Diode is OFF

$V_i \le -0.7 \text{ V}$ $V_o = \text{V in} + 0.7 \text{ V}$ Diode is ON

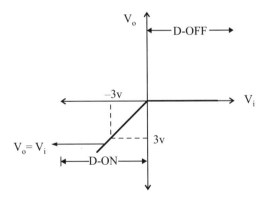

FIGURE 2.7 Practical diode of transfer characteristic

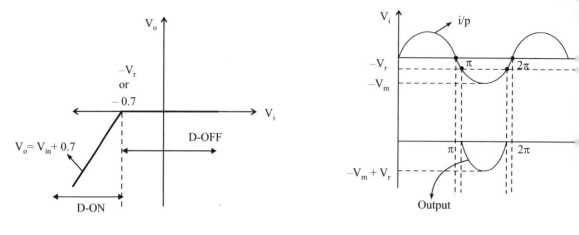

FIGURE 2.7(a) Practical diode of
Transform char clytus

FIGURE 2.8 Practical diode of input
and output waveform

Thus due to V_γ the part of Negative half cycle also get clipped off along with a positive Half cycle of the input in a positive series clipper circuit

2.5 CLIPPING ABOVE REFERENCE VOLTAGE V_R

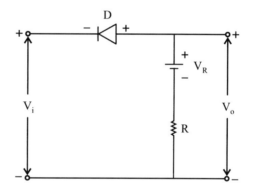

FIGURE 2.9 Clipping above V_R

Fig.2.9 shows clipping above V_R

When diode is forward biased D-ON condition. It acts as a short circuit. When diode i reverse biased diode is OFF condition. It acts as an open circuit.

Mathematically this can be written as

$V_i \le V_R$ $V_o = V_i$ Diode is ON

$V_i > V_R$ $V_o = V_R$ Diode is OFF

When V_i greater than V_R the diode is reverse biased

When V_i less than V_R the diode is forward biased

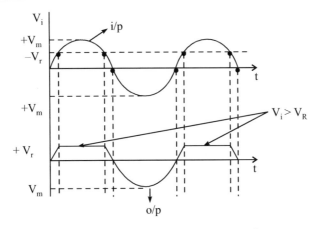

FIGURE 2.10 input and output waveform

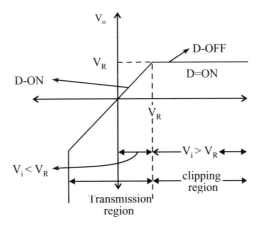

FIGURE 2.11 Transfer characteristics

2.6 CLIPPING BELOW REFERENCE VOLTAGE VR

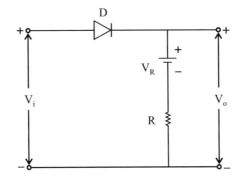

FIGURE 2.12 Clipping below VR

Mathematically written as

$V_i < V_R$ $\qquad$ $V_o = V_R$ $\qquad$ Diode is OFF

$V_i > V_R$ $\qquad$ $V_o = V_i$ $\qquad$ Diode is ON

The clipping below VR circuit as shown in Fig.12. This clipping circuit which clips the portion of waveform

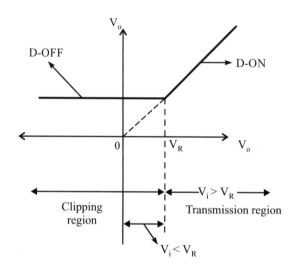

FIGURE 2.13 Transfer characteristics

The transfer characteristic are find out is shown in Figure 2.13.

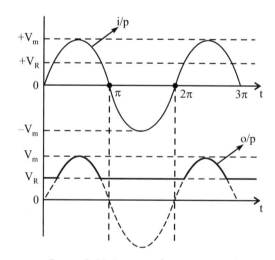

FIGURE 2.14 Input and output waveform

2.7 ADDITIONAL DC SUPPLY IN SERIES WITH DIODE

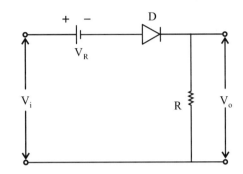

FIGURE 2.15 DC supply in series with diode

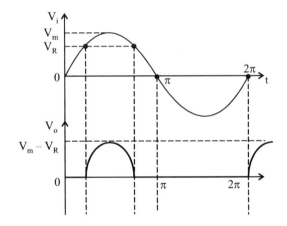

FIGURE 2.16 Input and output waveform

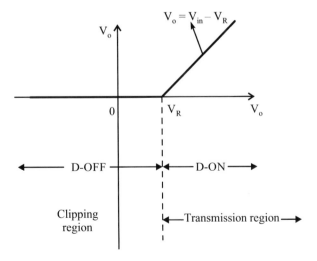

FIGURE **2.17** Transfer characteristic

Mathematically equation can be written as

$V_i > V$ $V_o = V_i - V$ Diode is OFF

$V_i \leq V$ $V_o = 0$ Diode is ON

The V_i voltage is not greater than V_R. The diode is not start conduction during positive Half cycle

$$V_o = V_i - V$$

During negative cycle the input voltage V_i is at its maximum V_m, the output will achieve its peak and will be equal to $V_m - V_R$. Diode, reverse biased cannot conduct at all. Hence the negative portion of the input will be clipped OFF. The output $V_o = 0$ input and output waveform as shown in Fig. 2.6

2.8 PARALLEL POSITIVE CLIPPER

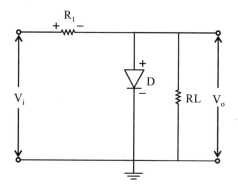

FIGURE **2.18** Parallel clipper of positive

In a parallel clipper circuit the diode is connected across the load terminal. It can be used to clip (or) limit the positive (or) negative part of the input signal as per the requirement as shown in fig 2.20.

The mathematical equation is written as

$V_i \geq 0$ $V_o = 0$

$V_i < 0$ $V_o = V_i \, R_L / R_1 + R_L$

During positive cycle of the input V_i, the diode - becomes forward biased.

When the diode is forward biased it acts as a short circuit. The current I flows through the diode (D). The diode across drop is short circuit is zero. In this condition positive half cycle gets clipped off.

During negative cycle of the input V_i the diode is become reverse biased. It acts as an open circuit, the outer current flows through the R_L.

Hence, $V_o = \dfrac{V_i R_L}{R_1 + R_L}$

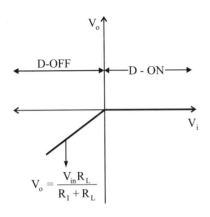

FIGURE 2.19 Transfer characteristics

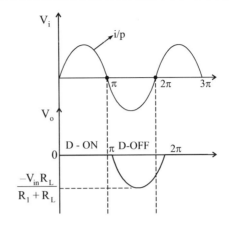

FIGURE 2.20 Input and output waveform

2.9 PARALLEL POSITIVE CLIPPER WITH CUTIN VOLTAGE DIODE

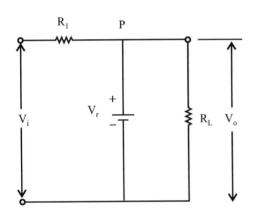

FIGURE 2.21 Parallel positive clipper with V_r

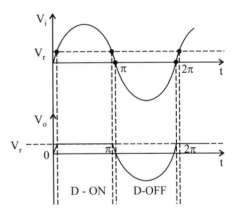

FIGURE 2.22 input and output waveform

In this circuit piecewise linear model of diode is used. When the potential of Node 'P' becomes just equal to the cut in voltage of diode, the diode start conducting $V_r = 0.7$ V of silicon, forward resistance r_f is $V_p = 0.7$V

Thus almost positive cycle gets clipped off.

During negative Half cycle V_i input voltage less than $V_\gamma(0.7$ V) then the diode become reverse biased. It act as an open circuited, current flows through the R_1 and R_L. The output given by

$$V_o = \frac{V_i R_L}{R_1 + R_L} = V_i$$

Mathematical equation is

$V_i \geq V_\gamma$	$V_o = V_\gamma$	Diode is ON
$V_i < V_\gamma$	$V_o = V_i$	Diode is OFF

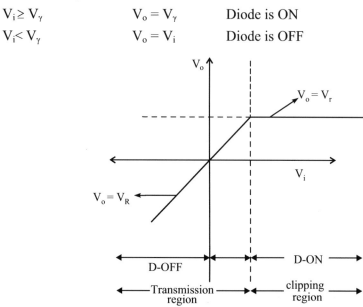

FIGURE 2.23 Transfer characteristics

2.10 PARALLEL NEGATIVE CLIPPER

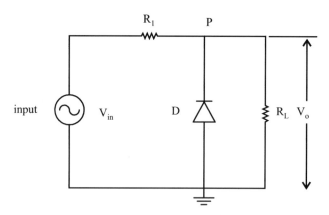

FIGURE 2.24 Parallel negative clipper

The mathematical equation can be written as

$V_i \geq -V_\gamma$ $\qquad V_o = V_i$ $\qquad$ Diode is ON

$V_i < -V_\gamma$ $V_o = -V_\gamma$ Diode is OFF

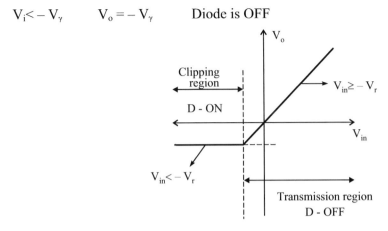

FIGURE 2.25 Transfer characteristics

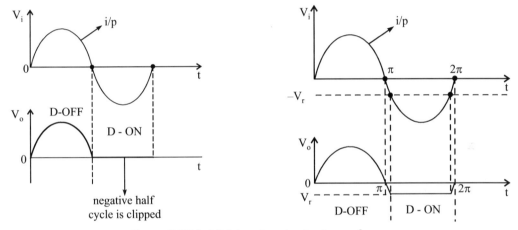

FIGURE 2.26 (a) (b) Input and output waveform

.11   PARALLEL CLIPPER WITH REFERENCE VOLTAGE V_R

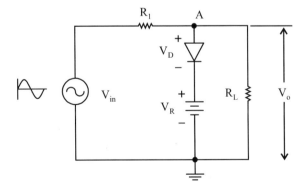

FIGURE 2.27   Parallel positive clipper with reference voltage

In Positive Cycle:

V_i is positive but less then V_R

When V_i is positive less than VR the diode is in OFF condition. It act as open circuited. Hence, the output voltage

$$V_o = \frac{V_i R_L}{R_1 + R_L}$$

$R_L >> R_1$

$V_o = V_i$

When V_i is positive greater than V_R

$V_i > V_R$ D turns into ON condition, it act as a short circuited. The output voltage i same as voltage of node A. Which is V_R as the drop across ideal ON diode is zero.

$$V_o = V_R$$

In Negative Cycle:

$V_i < V_R$ hence diode is in OFF condition. It acts as an open circuited

$$V_o = \frac{V_i R_L}{R_1 + R_L} = V_i \qquad\qquad R_L >> R_1$$

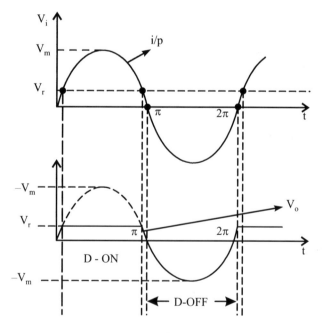

FIGURE 2.28 Input and output waveform

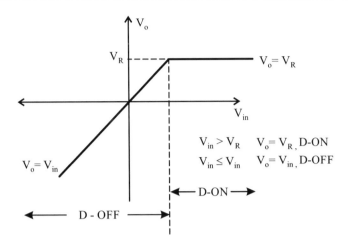

FIGURE 2.29 Transfer characteristics

$V_i > V_\gamma$, $V_o = V_R$ Diode is ON

$V_i \leq V_i$ $V_o = V_i$ Diode is OFF

2.12 PARALLEL NEGATIVE CLIPPER WITH VR (REFERENCE VOLTAGE)

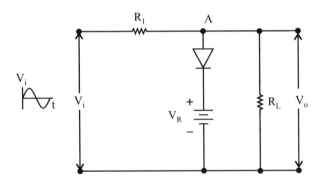

FIGURE 2.30 Parallel negative clipper with V_R

Assume Ideal Diode:

Positive cycle of the input when $V_i = 0$ V due to the diode turn in to ON condition, immediately node A is $A = V_R$ the diode is turn in to ON condition

$$V_o = A = V_R$$

When $V_{in} > V_R$, diode becomes OFF condition

$$V_o = \frac{V_i R_L}{R_1 + R_L} \square \; V_{in} \qquad R_L \gg R_1$$

Negative Cycle:

When mathematical equation can be written as, $V_i < VR$, the diode become ON condition

$V_o = V_R$

$V_{in} \leq V_R$ $V_o = V_R$ Diode is OFF

$V_{in} > V_R$ $V_o = V_{in}$ Diode is ON

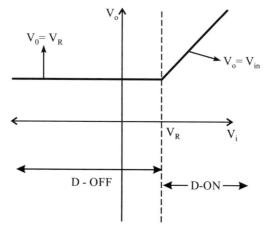

FIGURE 2.31 Transfer characteristics

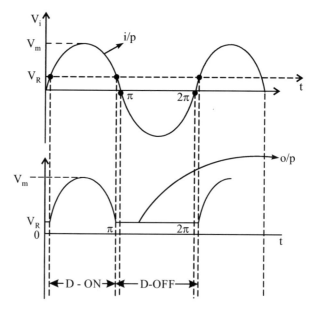

FIGURE 2.32 Input and output waveform

2.13   TWO WAY PARALLEL CLIPPER CIRCUIT

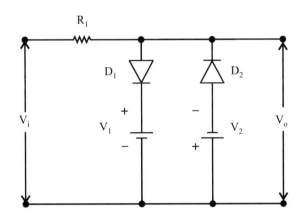

FIGURE 2.33 Two way parallel clipper

Input is purely sinusoidal signal

$$V_{in} = V_m \sin \omega t$$

Now D_1, D_2 diodes are idea diodes

During positive cycle

$$
\left.
\begin{array}{ll}
V_{in} < V_1 & V_o = V_{in} \\
V_{in} > V_1 & V_o = V_1
\end{array}
\right\} \quad \text{Positive cycle}
$$

$$
\left.
\begin{array}{ll}
V_{in} < V_2 & V_o = V_2 \\
V_{in} > V_1 & V_o = V_{in}
\end{array}
\right\} \quad \text{Negative cycle}
$$

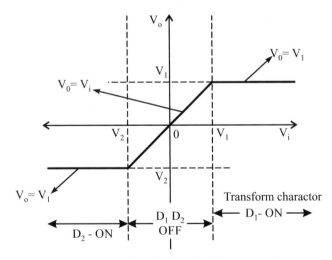

FIGURE 2.34 Transfer characteristics

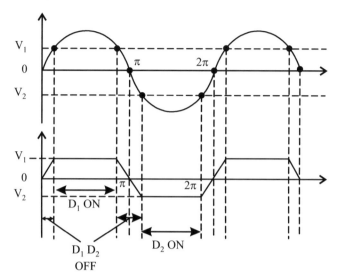

FIGURE 2.35 Input and output characteristics

2.14 TRANSISTOR CLIPPER

Clipping operation can also be performed with the help of transistors. It is possible to remove a portion of both extremities of an input waveform by driving a transfer into either cut-off (or) saturation. When the transitions cross from the cut off region into active region and second occur when the transfes crosses from the active region into saturation region. Therefore, if the peak to peak value of the input waveform is such that it can carry the transistor across the boundary between the cut off and active and saturation region, a portion of the input waveform will be clipped.

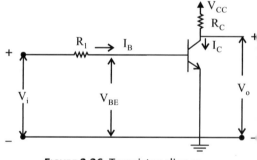

FIGURE 2.36 Transistor clipper

The transitor clipping circuits as shown in fig, the resistance R in the circuit represents either the source resistant R (or) the base resistant R_b in the base lead. This resistance R must be large in capariso with the input resistance of the transistor in active region. Assume that cut in voltage of the emitter base diode is negleble. The input base current will very nearly have the waveform of the input voltage because the base circuit is

$$ib = (V_i - V_\gamma)/R$$

$$\left. \begin{array}{l} V_\gamma = 0.1 \text{ Ge} \\ V_\gamma = 0.5 \text{ Si} \end{array} \right\} \quad V_{BE}\text{– cut in voltage}$$

Transistor clipper with ramp input:

Consider the ramp type of input which starts below cut off and increases till transister enter into saturation

For the transistor from b- parameter equivalent circuit can be writen

$$V_{BE} = ib.hfe$$

$$ib = \frac{V_i}{R + hie}$$

$$V_{BE} = \frac{V_i \, hie}{R + hie}$$

Above the equation differentiating both sides

$$\frac{dV_{BE}}{dt} = \frac{hie}{R + hie} \frac{dvi}{dt}$$

Above equation the relation between slop of the input and slop of the V BE graph

$$ib = \frac{Vi}{R + hie} \text{ this equation on differentiating both side}$$

$$\frac{dib}{dt} = \frac{1}{R + hie} \frac{dvi}{dt}$$

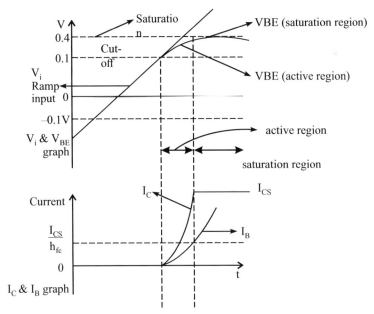

FIGURE 2.37 Input and output waveform

Fig (a) the step of the base current graph where is decreases, the transistor enter from cut off to active and the active to saturation region.

$$IC = \beta \, IB \text{ active region}$$

$$ie = \frac{VCC - VCE(sat)}{RC} \square\ I_{CS} \text{ (Constant)}$$

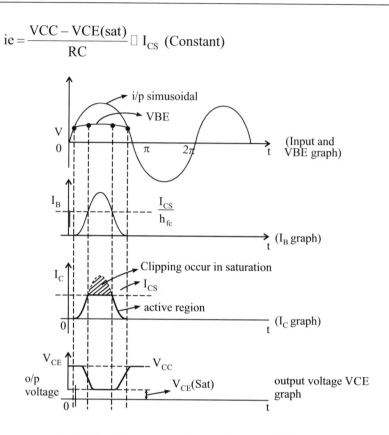

FIGURE 2.38 Transistor clipper with sinusoidal input

2.15 EMITTER COUPLED CLIPPER

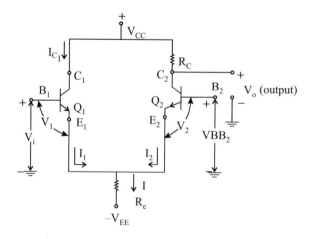

FIGURE 2.39 Shows the circuit diagram of emitter

It has two level clipper using transistors. The emitter coupled clipper is that circuit in which emitter of the transistor are coupled together. It has two clipping levels and hence it is called double ended clipper. The transistor Q_2 base is fixed at voltage VBB_2 and input is applied to B_1 and emitter of two transistor are coupled together and connected to $-VEE$ through common emitter resistance R_e.

The constant of voltage VBB_2 is supplied to the base of Q_2 such that, when Q_1 is OFF condition than Q_2 is in active region.

Totally V_i is the input voltage is negative which constant that Q_1 is cut off (OFF condition) at the time VBB_2 keeps Q_2 is in active region. When Q_1 is cut-off region only Q_2 carries the emitter current. As V_i id increases at some level, Q_1 becomes active region to cut off region. At the time both transistor will carry the currents. As both transitions are in active region the input waveform appear at the output. further V_i increases due to common emitter current. Emitter follows the base of Q_1 and emitter current increases. R_e across drop increases. Due to this emitter E_2 of Q_2 becomes more positive with respect to the base. Now VBB_2 is constant particular instant E_2 voltage become so positive enabling drives Q_2 to cut off from saturation. This causes second clipping of the output.

The emitter current I is given by

$$I = I_1 + I_2 = \frac{VEE + VBB_2 - V_2}{Re}$$

Applying KVL to base to emitter loop of Q_2 transistors

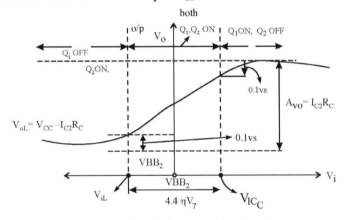

FIGURE 2.40 Transfer characteristics

When Q_1 is OFF $I_1 = 0$ $I_2 = I$

Lower level of output is given

$$V_oL = VCC - IC_2RC$$

When Q_2 is OFF $I_2 = 0$ $I_1 = I$

The upper level of output is given by

$$V_oC = VCC$$

Apply KVL V, V_2, and VBB$_2$

$$+VBB_2 - V_2 + V_1 - V_i = 0$$

$$V_i = VBB_2 + V_1 - V_2$$

Hence the change in output level from V_oL to V_oU is exponential. From the transisto
current and base emitter voltage it can be written as

$$V_i = VBB_2 + \eta V + ln \left[\frac{I_1}{I_2} \right]$$

V_{iu} corresponds

$$I_2 = 0.1I, \ I_1 = 0.9TI$$

V_{iL} corresponds to $I_1 = 0.1I$ and $I_2 = 0.9I$

$$Viu = VBB_2 + \eta V_T//ln \left[\frac{0.9I}{0.1I} \right] = VBB_2 + \eta V_T |n_9$$

$$ViL = VBBL + \eta V_T/n \left[\frac{0.1I}{0.9I} \right] = VBB_2 - \eta V_T |n_9$$

Total input voltage switch Δ Vi

$$\Delta Vi = Viu - ViL = 2\eta V_T /n9$$

$$\Delta Vi = 4.4\eta V_T$$

2.16 VOLTAGE COMPARATOR

The voltage comparator function is totally different from a clipping circuit. In
comparator circuit reproduction of a part of signal waveform is not any interest. Th
clipping circuits may also be used to perform the operation of comparator. A
comparator circuit is used to identify the instant at which the arbitrary inpu
waveforms attain a particular reference level.

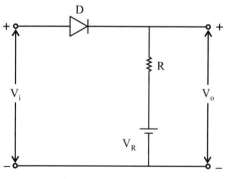

FIGURE 2.41 Diode comparator

$\Rightarrow$

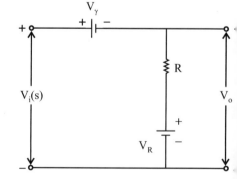

FIGURE 2.42 Equalent circuit

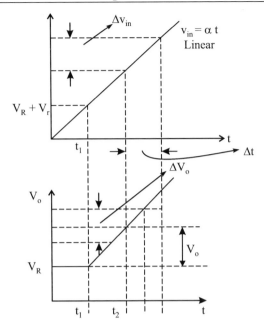

FIGURE 2.43 Input and output waveform

The output signal is taken as ramp waveform for convenience. At $t = t_1$, V_{in} becomes equal to $V_R + V_\gamma$, after where V_o increases along with the input signal V_{in}. The comparator output is given to a particular device, this device will respond. When comparator voltage increased to level of V_o above V_R.

When the input voltage Vi is less than reference voltage V_R, the diode D is ON condition and output is fixed at V_R. When Vi > V_R the Diode D is OFF condition and $V_o = V_i$ the break occur at $V_i = V_R$ at $t = t_1$.

Comparator may be Non-regenerative the (or) regenerative. Clipping circuits fall into the category of Non-regenerate comparator. In regenerative comparator, positive feed base is employed to obtain an infinite forward gain (unity Loop gain). Regenerative comparators examples are Schmitt trigger and blocking oscillators.

Applications of Voltage Comparator

(i) For accurate time measurements i.e., they are used as voltage to time convertor.

(ii) For converting sin wave into a series of pulse. It is called timing marker generator

(iii) Pulse time modulation is used in communication

(iv) In phase meter to measure the phase angle between $0°$ to $60°$

(v) To obtain square wave from since wave

(vi) In amplitude distribution analysis

(vii) In analog to digital converter (ADC). These are used to convert data from physical system to equivalent digital form

2.17 CLAMPER CIRCUITS

The circuit which is used to add to dc. Level as per the requirement to the a.c. output signal are called clamper circuits. These are used to clamp or fix the extremity of a periodic waveform to some constant reference level V_R. Under steady-state condition these circuits restrain the extremity of the waveform from going beyond V_R the capacitor, diode and resistor are the three basic elements of a clamper circuit. The clamper circuits are also called d.c. rest over (or) d.c inserter circuits, dc re inserter.

Depending upon this whether the positive d.c. (or) negative d.c. shift is introduced in the output waveform.

Classification of clamping circuits.

   (i) Negative clamper
   (ii) Positive clamper

2.17.1 NEGATIVE CLAMPER

In negative clamper as shown is Fig 2.44. In negative clamping the positive extremity of the waveform appears below the reference i.e., the output is negatively clamped with respect to the level.

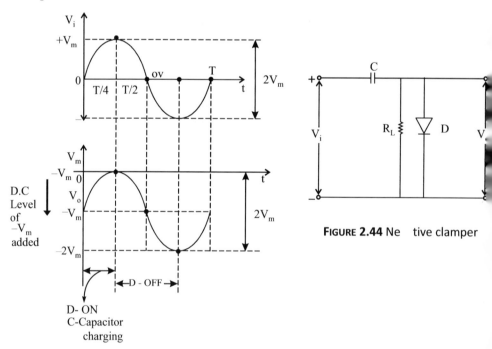

FIGURE **2.44** Ne tive clamper

FIGURE **2.45** input and output waveform

During the first Quarter of positive cycle of Vi input vo
the capacitor get charged through up to the maximum value of V_m of the input signal V_i. The input signal rises from zero to the maximum value is ideal diode. The voltage

across it a zero the capacitor C is changed through the series combination of the signal source and the diode and the voltage across C rises sinusoidally. At the end of the first Quarter cycle $V_c = V_m$.

In the negative half cycle of V_{in} the diode will remain reverse biased. The capacitor starts discharge through the R_L. At the time constant $R_L C$ is very large, it can be approximated that the capacitor holds all its charge and remains charged to V_m, during two periods also. Summary as

$$V_o = V_{in} - V_c = V_{in} - V_m \text{ (negative cycle)}$$

$$V_o = - V_m \qquad\qquad \text{for } V_i = 0$$

$$V_o = 0 \qquad\qquad\qquad \text{for } V_i = V_m$$

$$V_o = - 2V_m \qquad\qquad \text{for } V_i = - V_m$$

.17.2 POSITIVE CLAMPER

In positive clamping, the negative extremity of the waveform is fixed at the reference level and the another waveform appear above the reference level. The output waveform is positively clamped with reference to the reference level.

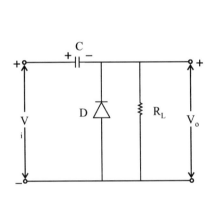

ltage, diode is forward biased,

FIGURE 2.46 positive clamper

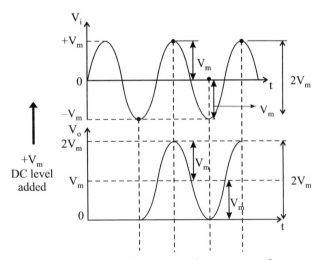

FIGURE 2.47 input and output waveform

In the negative cycle:

When V_i input voltage is negatively applied the diode D gets forward biased condition and in a few cycle the capacitor gets charged to V_m. Under steady state condition the capacitor acts as a constant voltage source and the output is

$$V_o = V_i - (V_m) = V_i + V_m$$

In the positive cycle:

The diode reverse biased. The capacitor starts discharge through R_L. But due to large time constant. It get discharged during positive cycle of V_{in}.

$$V_o = V_{in} + V_m$$

$$V_o = V_m \qquad\qquad \text{for } V_i = 0$$

$$V_o = 2V_m \qquad\qquad \text{for } V_i = V_m$$

$$V_o = 0 \qquad\qquad \text{for } V_{in} = -V_m$$

2.18 CLAMPING CIRCUIT THEOREM

The clamping theorem is related to the area of under output waveform in forward direction diode is in ON condition and the area under output waveform in reverse direction diode is in OFF condition under steady state condition.

Statement: The clamping circuit theorem states that under steady state condition for any input waveform, the ratio of the area under the output curve in forward direction (when diode is conducts) to the area under the output circuit in reverse direction (when diode is not conduct) is given by

$$\frac{A_f}{A_r} = \frac{R_f}{R_r}$$

A_R = Area under output circuit in reverse direction

A_f = Area under output circuit in forward direction

R_f = Forward resistance of the diode

R_r = shunt resistance

Proof of Clamping Circuit Theorem

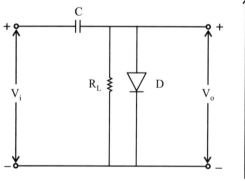

FIGURE 2.48 Clamper circuit

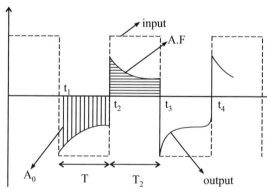

FIGURE 2.49 Input and output waveform

In these clamper circuit input square wave is applied

During period T, i.e., from t_1 to t_2

Diode is OFF condition hence capacitor is discharge

Let iR be the discharge capacitor current

The charge lost in the internal $t_1 - t_2$ is given by

$$Q_D = \int_{t_1}^{t_2} iR dt \qquad iR = \frac{VR}{R}$$

VR = reverse output voltage

$$Q_D = \int_{t_1}^{t_2} \frac{VR}{R} dt = \frac{1}{R}\int_{t_1}^{t_2} VR dt$$

In the interval t_2 to t_3 the diode is ON condition. Thus it is the charging current of diode. The charge gained is

$$Q_C = \int_{t_2}^{t_3} i_f dt$$

$$if = \frac{V_f}{R_f} \quad R_f = \text{forward reactance}$$

$$Q_C = \int_{t_2}^{t_3} \frac{Vt}{Rd} dt = \frac{1}{R_f}\int_{t_2}^{t_3} V_f dt$$

Under steady state condition charge lost must be equal to charge gained i.e., $Q_A = Q_C$

$$\frac{1}{R}\int_{t_1}^{t_2} V_R dt = \frac{1}{R_f}\int_{t_2}^{t_3} V_f dt$$

$$\int_{t_1}^{t_2} V_R dt = \text{Area under output for reverse direction when diode is OFF} = AR$$

$$\int_{t_2}^{t_3} V_f dt = \text{Area under output for forward direction when diode is ON conduction} = AF$$

$$\frac{A_R}{R} = \frac{A_f}{R_f}$$

$$\frac{A_f}{A_r} = \frac{R_f}{R}$$

2.19 CLIPPER CIRCUITS AND WAVEFORM

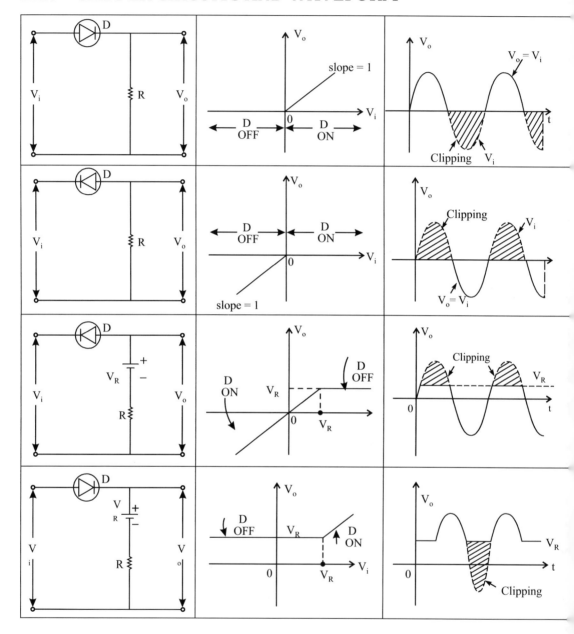

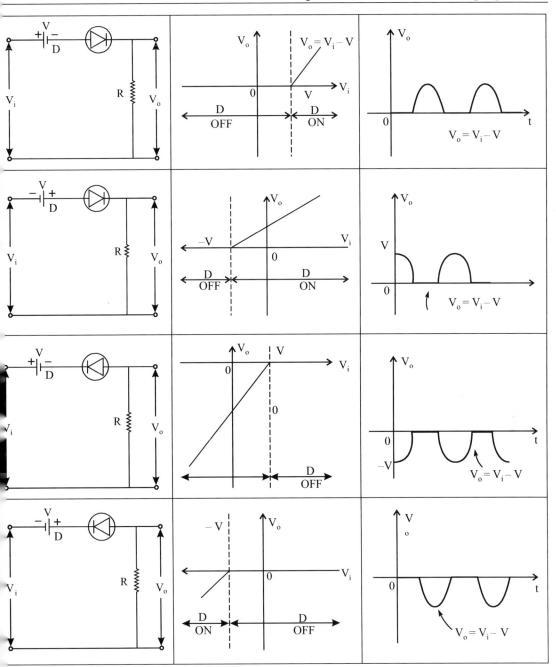

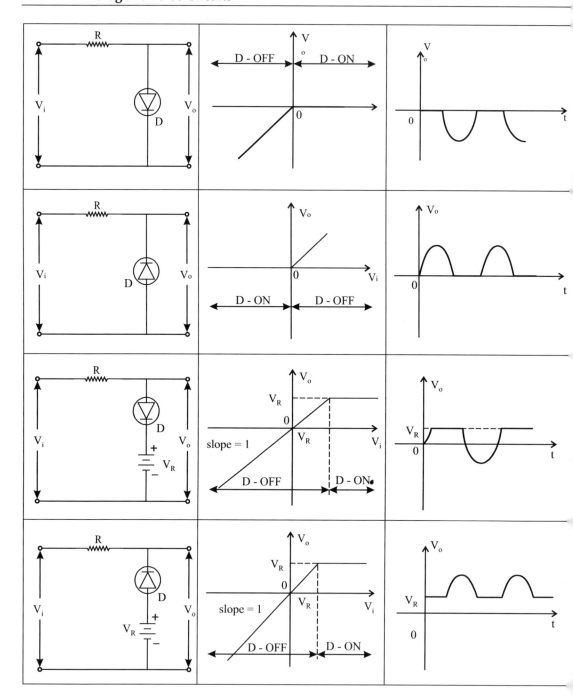

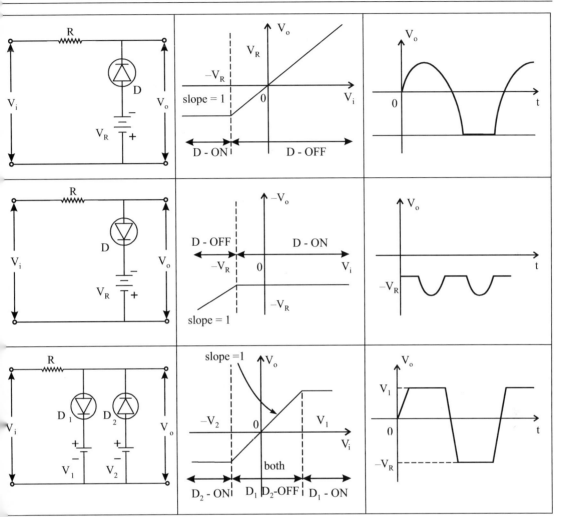

3

Switching Characteristics of Devices

3.1 INTRODUCTION

Most of the semiconductor devices in wave shaping and switching circuit function as electric switches. Several semi conductor devices can be operated as electric switches as they have two functional states ON and OFF. In practical applications, it is necessary to use the Junction diodes, transistor, vacuum tubes, thermionic diodes when diode is forward biased, it acts as a closed switches as its forward resistance is zero. Similarly when diode is reversed biased it act as an open switch as its reverse resistance is very high. When transistor is cut off, it acts as open circuit switch OFF condition. When transistor is saturated it acts a closed circuit switch ON condition. The understanding gained by the study of the physical behaviour of diode and transistor is applied in the design of a transistor switch.

3.2 SWITCHING CHARACTERISTICS OF JUNCTION DIODES

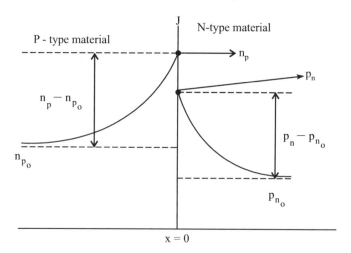

Figure 3.1 Forward biased P.N. Junction

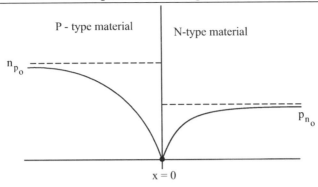

Figure 3.2 Reverse biased P.N. junction

Minority charge concentration across the PN junction when the PN junction diode is switched from ON condition to OFF condition state. It takes finite time to attain a steady state. The time consists of interval and transient tince before the diode attains a steady sate. The behaviour of the PN junction diode during the time is called switching characteristic of diode.

When forward biased (ON condition) large number of electrons move to n side diffusing to P sides. Similarly large number of hole diffusing into N side from P side.

When diode is forward biased (ON condition) npo is carrier concentration of electron on P side at thermal equilibrium and Pno is carrier concentration of hole on the N side of the thermal equilibrium. This concentration level far away from the junction. It increased towards the junction as shown in the figure (3.1)

When the PN junction diode is reverse biased again far from the junction, the minority charge carrier concentration is npo on P side and pno is N side. Hence minority carrier concentration decreases to zero at the junction steady state as shown in figure 3.2 until such time as the injected (or) excess minority carrier density pn - pno (or) np - npo drops normally to zero. The PN junction diode will continue to conduct easily and current will be determined by the external resistance in the diode circuit.

3.3 SWITCHING TIMES OF PN JUNCTION DIODE

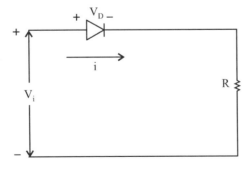

Figure 3.3 The diode circuit with R

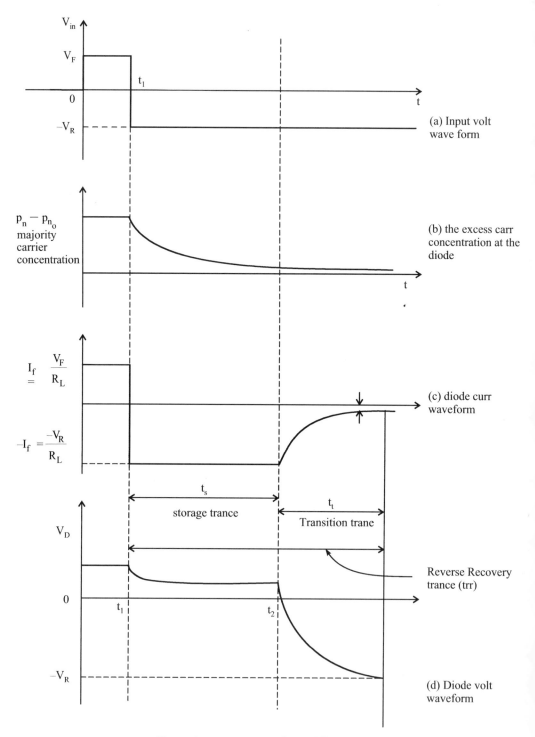

Figure 3.4 output waveform different parameter

To understand the various switching times, consider simple diode circuit and input waveform fig 3.4(a)

Up to $t = t_1$; $V_i = V_F$ the value of R resistance is large the drop across R is large compared with the drop across the diode.

$$I_F = \frac{VF}{R}$$

At the time t_1 the applied voltage suddenly reverse biased, the reverse voltage is – VR is applied to the circuit

$$V_i = -VR \quad -IR = -\frac{VR}{R} \text{ until the } t = t_2 \text{ as shown Fig 3.3}$$

The number of minority carrier later time to reduce from pn – pno to zero as shown in fig 3.4(b) the minority carrier density reaches its equilibrium state. Due to this at t_1 current junction reverse biased and remains at that reversed value – I_R till the minority carrier concentration reduces to zero

$$-IR = -\frac{VR}{R}$$

this condition to flow till time t_2

Storage time: During time t_1 and t_2, the minority charge carrier remains storted and decreases slowly to zero, denoted as t_s.

From t_2 onwards the diode voltage starts reverse and diode current starts decreasing as shown in figure 3.4(c) at $t = t_3$ the PN junction Diode completes reverse biased condition to attain a steady state.

Transistor (or) Transition Interval: The time from t_2 to t_3 ie., time required by the PN junction Diode current to reduce to its reverse saturation value is called the transition time. The PN Junction Diode total time required is the sum of storage time and transition time to recover completely from the charge of state is called reverse recovery time of the PN Junction Diode. It is denoted by t_{rr}. This a high speed switching application for Quick switching from ON to OFF state

$$T_{rr} = t_s + t_e$$

The reverse recovery time depends on the RC time constant

3.4 BREAKDOWN MECHANISMS IN PN JUNCTION DIODE

When the reverse biased voltage is applied to the PN junction Diode, it reaches the breakdown value VB and an increase in the reverse current diode sinusoidal rice attributed to two kinds of breakdown mechanism. The diode no longer blocks current and the Diode current can now be controlled only by the resistance of the external circuit.

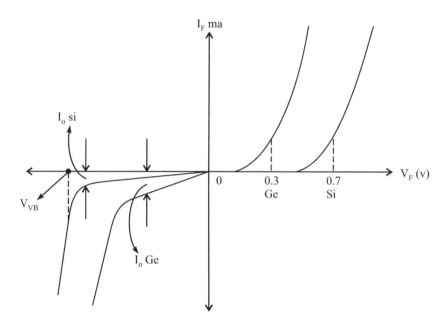

Figure 3.5 Volt Ampere characteristic of breakdown in PN junction diode

3.4.1 ZENER BREAKDOWN

In zener breakdown the electric field inside the semi conductor PN Junction Diode becomes very large. This strong electric field pulls out electrons from their covalent bonds. The dived disruptions of the bonds create a large number of holes and electrons. Initially carriers available do not gain enough energy to disrupt bonds. It is possible to initiate breakdown through a direct rupture of the bonds because of the existence of strong electron field. This breakdown is called as "Zener Breakdown". Zener breakdown occur at voltage below 6V. Zener diode is negative temperature coefficient. i.e., temperature increases as the breakdown voltage decreases.

3.4.2 AVALANCHE BREAKDOWN

In avalanche multiplication the thermally generated minority carriers are accelerated by the external electric field across the PN Junction. The thermally generated minority carrier thus gains large kinetic energy. When these charge carrier acquire sufficient kinetic energy, they collide with the valance electronics in the covalent bond shell success full in dislodging leading to the creation of secondary electrics. Thus the secondary electrons also succeed in dislodging electrons from the valance shell. This kind of cumulative increase of charge carrier is termed as Avalanche multiplication. The avalanche breakdown diode exhibits find out is positive temperature coefficient i.e., the breakdown voltage increases with variance in temperature.

.5   TRANSISTORS AS A SWITCH

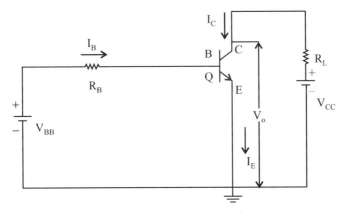

Figure 3.6 (a) Transistor as switch

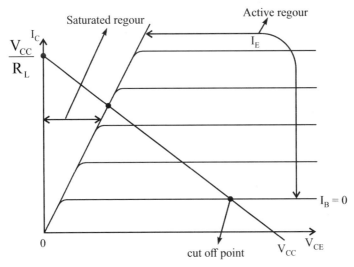

Figure 3.7 CE configuration output characteristics

Transistor transfers the signal from low resistance to high resistance. Bipolar junction transistor consist of three terminal emitter, base, collector and two junctions. They are emitter base junction and collector base junction. The transistor can be used as switch. When emitter-Base junction and collector base junction are reverse biased, the transistor treated as cut – off region. The application is transistor OFF condition it acts as an open switch. When both junctions are forward biased it operate saturation region. The application is switch ON condition, it acts as a closed switch. When the transistor is saturation regions the both junction voltages are very small but operating currents are large. When both junction are cut-off region the junctions voltage are very large and the currents are very small (small leakage current). The basic transition used in switching application is called as inverter.

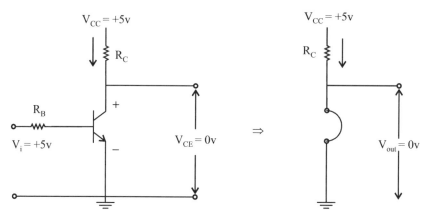

Figure 3.8 Transistor saturation

Figure 3.9 Transistor Switch ON (closed)

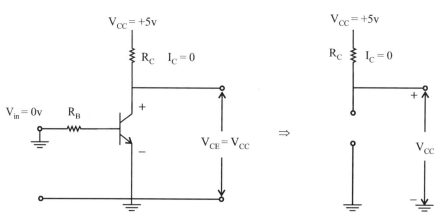

Figure 3.10 Transistor cut-off

Figure 3.11 Transistor OFF (open)

If IC and IB can be determined independently from the circuit under consideration, then the transistor is in saturation IR ≥ IC/β. If VCB is determined from the circuit configuration and this quantity is negative for a n-p-n transistor (or positive for p-n-p transistor) then the transistor is in saturation.

In saturation conductor voltage between emitter and collector VCE saturated, is typically 0.2 V to 0.3 V. The collector current at the saturation is called IC (sat) and nearly equal to the value VCC/RC, neglecting VCE(sat).

When the transistor is saturated it is said to be ON, the analysis has indicated that high input to the inverter [+ 5v] results in a low output voltage (≈ 0 V). Since in saturation, VCE ≈ 0 V and V out is equal to VCE. When transistor is ON VCE ≈ closed switch and the current is maximum VCC/RC. When the transistor is cut off no current flows between collector and emitter and the voltage is maximum. It acts as a open switch.

3.6   TRANSISTOR BREAKDOWN VOLTAGE

When transistor is used as switch which occurs at the collector with switching is approximately equal to VCC. Since this voltage is used to operate other circuits and devices for the sake of reliability of operation, VCC should be made large as possible. The maximum allowable VCC voltage depends not only on the characteristics of the transistor but also on the associated transistor base.

The collector base junction is reverse biased, the reverse biased voltage increases with emitter open circuited. In a common base, configuration causes the A valances multiplication of the Ico that crosses the collector junction the current becomes MIco, where M is the factor by which the original current ICo is multiplied by the avalanche effect. The reverse voltage is high at which multiplying factor M becomes infinite is known as B V_{CBO}. At the voltage, breakdown occurs.

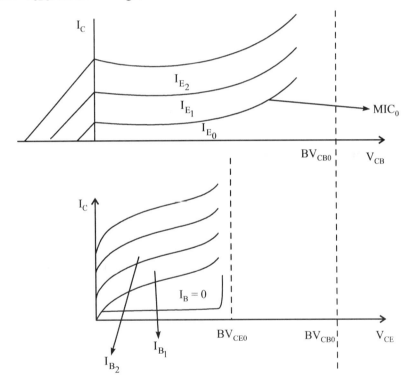

Figure 3.12 (a) Common base characteristic extended into break down region **(b)** Common emitter characteristic into breakdown region

In the CE configuration indicates avalanches breakdown that the collector to emitter the breakdown voltage with open circuited base represented as BV_{CEO}. Usually B_{VCEO} is Half of BV_{CBO} .

In avalanche multiplication factor M depend on two voltage V_{CB} between collector and base. It is given by

$$M = \frac{1}{1-\left[\dfrac{VCB}{BV_{CBO}}\right]^n}$$

In the above equation, parameter n large from 2 to 10 control the sharpness of the onset breakdown. When 'n' is large M continues at nearly unity until VCB approached very close to BV_{CBO} at which point M soars upwards abruptly. When n is small, the onset of breakdown is more gradual.

Fig 3.13(b) shows the common base configuration characteristics. It is extended into the breakdown region the circle for $I_E= 0$ is plot, as function if VCB of t the product of the reverse collector current I_{co} and the avalanche multiplication factor M. The abrupt growth in IC and BV_{CBO} is approached and there is a slow increase in IC over the active region.

I_E current flows through the emitter function a fraction of circuit αIE, where α is the common base configuration of current amplication factor (AI). Consider the avalanche A valance effect into account IC has magnitude M αIE. Due to the avalanche effect the common base current gain appear to α = Mα

The CE configuration

Since we have hfe $= \dfrac{\alpha}{1-\alpha}$, the CE current gain the conducting avalanche multiplication.

$$h*_{fe} = \frac{\alpha*}{1-\alpha*} = \frac{M\alpha}{1-M\alpha}$$

α is positive member and it is less than unity but very near to unity in magnitude Mα may equal unity in magnitude. When Mα = 1 any base current, no matter how small will give rise to an arbitrarily large collector current. It mean break down occurres when Mα = 1 at the voltage V_{CB} satisfying the equation

$$M = \frac{1}{\alpha} = \frac{1}{1-\left[\dfrac{VCB}{BV_{CBO}}\right]^n}$$

$$VCB = BV_{CBO} \sqrt[n]{1-\alpha}$$

Since VCB at breakdown is much larger than the small forward base to emitter voltage VBE. We may replace VCB by VCE in the equation

$$\frac{\alpha}{1-\alpha} = hfe$$

$$1 - \alpha = \frac{hfe}{\alpha} = \frac{1}{hfe}$$

∴   α is very close to 1

Substituting the value of $1 - \alpha$ in equation

$$BV_{CEO} = BV_{CBO} \sqrt[n]{\frac{1}{hfe}}$$

The Breakdown Voltage with Base not Open Circuited

Fig 3.15 the CE configuration with the base to the emitter through a resister RB. The breakdown voltage for circuit as BV_{CER}. probably BV_{CER} will lie between BV_{CER} and BV_{CBO}. To estimate BV_{CEO}, V_r is CE – 0.2 V and i→ 0.6. The collect current will flow entirely to the base and hence through RB, once threshold voltage is exceeded all the additional collector current will flow through the emitter junction and the corresponding breakdown voltage is BV_{CBO}. Therefore, when collector to emitter voltage is larger than BV_{CBO} and the cut in voltage of the emitter junction is reached, break down will occur when

$V_r = MI_{co}RB$

$$M = \frac{I_{Co}RB}{V_r}$$

$$M = 1/\alpha \qquad BV_{CER} = BV_{CBO} \sqrt[n]{\frac{1 - I_{co}RB}{V_\gamma}}$$

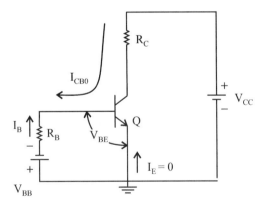

Figure 3.13 Base circuit

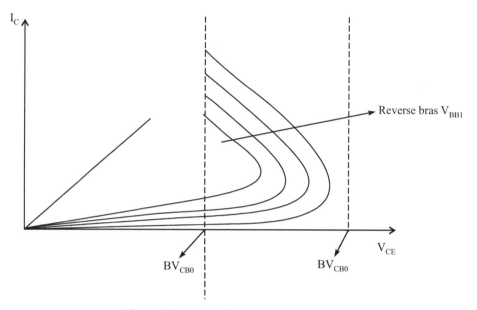

Figure 3.14 Break down characteristics

The above equation is valid only when the current through RB is very large in comparision with the emitter function current. If RB is large this condition satisfie then above equation is not applicable.

If the base is short-circuit to emitter i.e, when RB = 0

$$BV_{CER} = BV_{CBO} = BV_{CES}$$

Consider rbb total base resistance RB + rbb' $\therefore RB = 0$

$$BV_{CER} = BV_{CBR} \sqrt[n]{\frac{1 - I_{co}rbb'}{V_{\gamma}}}$$

The breakdown voltage may also be increased by restively the resistance RB to voltage VBB as shown fig 3.15(b). This introduces the back bias and the onse breakdown occurs when

$$MIC_o (RB + rbb') = V_r + V_{BB}$$

$$M = \frac{V_r + VRB}{IC_o (RB + rbb')}$$

Such breakdown is denoted $BV_{CER} \times 1$

$$M = 1/\alpha$$

$$BV_{CE\alpha} = BV_{CBO} \sqrt[n]{\frac{1 - I_{co}(RB + rbb')}{V_r + VBB}}$$

Design of Transistor Switch

The transistor acts as a switch is driven between saturation and cut off regions

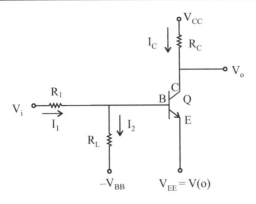

Design 1: when $V_i = V(o)$ (transistor is cut-off region OFF condition)

Transistor is open circuited V_B = Base Voltage

$$VB = -VBB\frac{R_1}{R_1+R_2} + V(o)\frac{R_L}{(R_1+R_L)}$$

If $V(o) = OV$ than $VB = -VBB\frac{R_1}{R_1+R_L}$

This voltage should be less than VBE cut off region for the transistor i.e., OV.

Design 2: When input $V_i = V(1)$ (transistor is saturation region Traction ON condition)

$$RC = \frac{VCC - VCE(sat)}{IC}, \qquad IC \Rightarrow \frac{VCC - VCE(sat)}{RC}$$

$$I_{B(min)} = \frac{I_C}{hfe(main)}$$

The current through R_1 resistance is given by

$$I_1 = \frac{V(1) - VBEsat}{R_1}$$

The current through R_2 resistance is given by

$$I_2 = \frac{VBE(sat) - (VBB)}{R_2}$$

$$I_B = I_1 - I_2$$

$$I_B = \frac{V(1) - VBEsat}{R_1} - \frac{VBE(sat) - (-VBB)}{R_2}$$

The current must be equal to $\dfrac{IC}{hfE}$

$$\dfrac{IC}{hfe} = \dfrac{VCC - VCE(sat)}{RChfe}$$

Assuming the value of VBE(sat) and VCE(sat) select such that

$$RC \Rightarrow \dfrac{VCC - VCEsat}{IC}$$

$$VB = VBB\dfrac{R_1}{R_1 + R_2} + V(o)\dfrac{R_L}{R_1 + R_L}$$

3.7 TRANSISTOR SWITCHING TIMES

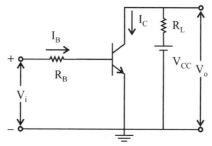

Figure 3.15 CE configuration

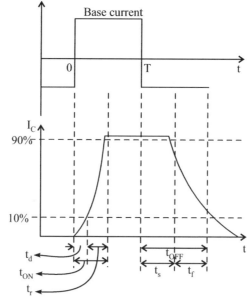

Figure 3.16 Output and input waveform

When the transistor is used as a switch it is consider saturation region (ON condition) and cut-off region (OFF condition). When the base input current is applied the transistor does, not switch on immediately. This is because of the junction capacitance and the transition time at electronics across the junction. The input waveform V_i is applied between base junction and emitter through resistance RB. The out waveform as shown in figure 3.16. In the time between input pulse and collator current flow is termed as delay time and required for IC. 90% if its maximum from 10% level is called "rise time" t_r. This is the turn ON time is the addition of t_r and t_d ($t_{ON} = t_d + t_r$).

When the input signal returns to initial position the collector current does not go to zero level immediately. If zero level is attained after turn off time, which is the sum of voltage time ts and fall time t_f. The fall time is required for IC to go from 90% to 10% of maximum level.

When a transistor is in saturation region. This has excesses of minority carrier stored in the base. The transistor can't respond until this saturation excesses charge has been removed.

Due to this a finite time elapses between the transistor of the input waveform and the time when collector current has dropped to 90% of IC (sat) and it is called as storage time (t_s).

4

Multi-vibrator

4.1 INTRODUCTION

The electronic circuits which are used to generate non-sinusoidal waveforms are called multi-vibrator. Multi-vibrator means many oscillations.

The multi-vibrator is nothing but two states operating in two modes. The modes are called state of the multi-vibrator.

Basically these are three types of multi-vibrators

1. Bistable Multi-vibrator
2. Monostable Multi-vibrator
3. Astable Multi-vibrator

Bistable multi-vibrator is one of the most useful circuits employed in all digital systems. The multi-vibrators are primary employed for two purposes: counting and storing. The Bistable multi-vibrator has two stable states the multi-vibrator can exist indefinitely in either of two stable states. It require an external pulse trigger to change from one stable to another these vibrating are also called Eccles Jordon circuit, target circuit, scale of-2-toggle circuit, flip flap and binary.

The monostable multi-vibrator has one stable state and one quasi-stable state. The quasi-stable states exist only for a finite period of time. When an external triggering pulse is applied to the circuit, the circuit goes into the quasi-stable state from its normal stable state After some time interval, the circuit automatically return to its stable state. Mono stable multi-vibrator are also termed as, one short, single shot, single cycle delay circuit, pulse structure, multi-vibrator gating circuit, a single swing. The multi-vibrator are used to voltage to time converter.

Astable multi-vibrator both states are quasi states. The astable multi-vibrator make successive transistor from one Quasi stable state to another Quasi stable state without any an external triggering signal. These multi-vibrators does not require any external pulse for the transistor is called recanting multi-vibrator. This multi-vibrator is used t voltage to frequency convertor.

State:

A multi-vibrator can be either an ON-OFF state or OFF - ON. It has two states – a multi-vibrate as a relative condition of the two transistors of the multi-vibrate that a condition of the first transistor and cut-off of the second transistor.

Stable State:

If the multi-vibrator can remain indefinitely in a state until the circuit is triggered by an external signal that state is known as a stable state.

Quasi stable state:

If the multi-vibrator can remain only for predetermined duration in a state and switches into the other state without the application of any external trigger is known as Quasi-stable state.

4.2 BISTABLE MULTI-VIBRATOR

A Bistable multi-vibrator has two stable states. There are two types of bistable multi-vibrator

1.   Collector coupled Bistable multi-vibrator
2.   Emitter coupled Bistable multi-vibrator

There are two types of collector-coupled bistable multi-vibrator

1.   Fixed bias multi-vibrator
2.   Self bias multi-vibrator

4.2.1   FIXED BIAS TRANSISTOR BISTABLE MULTI-VIBARTOR

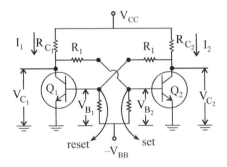

Figure 4.1 Shows the circuit diagram of fixed bias bistable

Multi-vibrator using transistor. The circuit uses two n-p-n transistors Q_1 and Q_2. The collector of Q_2 is coupled to the base of Q_1 through resistance R_1, while the collector of Q_1 is connected to the base of Q_2 through another resistance R_1. In one of the stable state, transistor Q_1 is ON (saturation) and Q_2 is OFF (cut-off) and other stable state Q_1 is OFF and Q_2 is ON, even though the circuit is symmetrical.

It is not possible for the circuit to remain in a stable state when both transistors are conducting (both transistor in above region) simultaneously and carry equal currents. It assume that both transistors are biased equally and carry equal currents I_1 and I_2 and there is a minute fluctuations in the current I_1 – let us say it increases in small amount – then the (Vc_1) voltage at the collector of Q_1 decreases. This results in decreasing in the base (vb_2) of Q_2 transistor so the Q_2 transistor in cut-off region does not conducted. I_2 decrease, the Q_2 transistor collector voltage increases (Vc_2). This result in an increase in the base of the Q_1 transistor (vb_1). So the current I_1 keep on increasing and I_2 current keeps on decreasing till the Q_1 goes into saturation and Q_2 goes into cut-off.

This is a stable state of the multi-vibrator. The circuit remain in the state till an external trigger pulse is applied at the set (or) reset terminal. If a positive going pulse is applied at the set (or) reset terminal, it will drive the Q_1 to cut-off and the transistor Q_2 to saturation. This is nothing but the second stable state of multi-vibrator. The circuit remains in this stable state through the applied pulse is removed.

$$I_1 > I_2$$

$$Q_1 - ON, \qquad Q_2 - OFF$$

Triggered pulse is applied $\qquad Q_1 - OFF, \qquad Q_2 - ON$

Set (or) Reset:

The value of R_1, R_2 and VBB must be selected such that in one stable state the base current is large enough to drive the transistor into saturation. Whereas in the second stable state the emitter function must be below cut-off. The change in the collector voltage resulting in to the saturation from one state to other is called output voltage swing denoted as $V\omega$. R_1 can be neglected.

$$V\omega = V_{c1} - V_{c2}$$

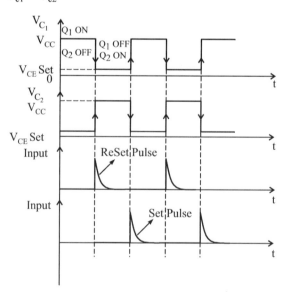

FIGURE 4.2 Output waveform of bistable multi-vibrator

.3 COLLECTOR CATCHING DIODES USING A FIXED-BIAS BINARY MULTI-VIBARTOR

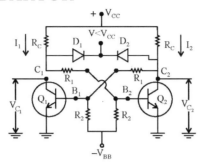

FIGURE 4.3 Collector catching diodes using fixed bias binary multi-vibrator

The collector voltage of a bistable multi-vibrator is used to drive other circuit. In such cases loading of multi-vibrator due to the circuit be considered. Such loading reduce the output voltage switching due to the decrease in the collector voltage.

Let Q_1 is cut-off (OFF condition) and Q_2 is saturation (ON condition). For the stable state of the multi-vibrator A-reduced V_{c1} will decreases IB_2 and Q_2 transistor may not be driven into the saturation. Hence the flip flap circuit components must be chosen such One transistor remains in ON while the other is OFF.

When the loading varies as the transistor operates, the constant output voltage swing and constant base saturation current required to maintain saturation of one transistor can be obtained by using two diodes – the two diodes are collector catching diodes and the auxiliary supply voltage V used in less than V_{CC} ($V < V_{CC}$).

When Q_1 in cut-off region the collector voltage V_{C1} increased to V_{CC}. When to becomes equal to V which is less than V_{CC}, the diode D_1 conducts; thus it act as a short circuit V_{C1} = V. The diode clamps the collector voltage V_{C1} at V. By saturating, circuit does not get loaded and I_{b2} is maintained at such a level that the transistor Q_2 is driven into saturation.

.4 SELF BIASED TRANSISTOR BISTABLE MULTI-VIBRATOR

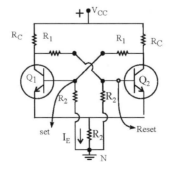

FIGURE 4.4 Self biased transistor at bistable multi-vibrator

In fixed bias bistable multi-vibrator circuit, two power supplies V_{CC} and $-V_{BB}$. Introduce a common emitter resistance R_e into the bistable multi-vibrator circuit as shown in the Fig.4.4

The binary using the common emitter resistor R_E to provide self biase is called self biased binary. The bistable multi-vibrator has two stable states – ON-OFF and OFF-ON. In both these states the drop across R_e remain almost same. When the circuit is going through a transistor from one state to other state the emitter current I_E change by ΔI_E. Assume that Q_1 cut-off and Q_2 in saturation. Hence to keep the voltage V_{EN} constant which provides the required self bias during the transition period.

4.5 SPEED-UP CAPACITOR (OR) COMMUTATING CAPACITOR

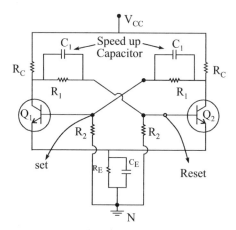

FIGURE 4.5 Practical self biased bistable multi-vibrator

The bistable multi-vibrator remains in the stable state till the triggering pulse is applied to set (or) reset terminal. When triggering signal is applied, conduction has to transfer, from one device to another. The transistor time means the interval during in which condition transfer from one transistor to another. Transistor time should be as small as possible. In practice, the switching characteristics can be improved by passing the high frequency components of the pulse. For this purpose small capacitance is used in sheet with the coupling resistor R_1; this results the transistor time to reduce considerably without disturbing the stable states because the base of the transistor makes the $R_1 - R_2$ attenuate

acts as un-compensated attenuator and so it will have a finite rise time $t_r = \left[\dfrac{R_1 + R_2}{R_1 + R_2}\right] C$

Since C_1 capacitance is to increase the speed of operator of the devices they are called speed-up capacitor (or) transpose (or) commutating capacitor. The speed up capacitance has the removal of charge stored at the base of the ON transistor due to minority carriers

If the speed-up capacitor have large the $R_1 - R_2$ network acts as on over composited attenuator; large value of capacitor have some disadvantages.

The smallest allowable interval between triggering is called resolving time of the bistable, that means the resolving time should be sufficient. So that, all the transients die out completely and the flip flop can be triggered reliably. The resolving time decides the maximum frequency of triggering to which biased can respond.

$$f_{max} = \frac{1}{resolving\ time}$$

$$f_{max} = \frac{1}{2C_1(R_1 \square R_L)} = \frac{(R_1 + R_L)}{2C_1 R_1 R_L}$$

The voltage across the speed-up capacitor C_1 need not charge during this transformer condition. After this transformer of condition, the capacitor are allowed to inter change their voltage. This additional time required for the purpose of completing the recharging of capacitor after the transformer of condition is called the settling time. The sum at transistor time and settling time is called "resolution time". Speed-up capacitor C_1 is small it gives transitions time increases and settling time will be small.

Improving Resolutions

The Resolution time for the binary can be improved by

1. Reducing the value of resistance R_1, R_2 , R_c
2. Reducing the stay capacitance
3. By not allowing the transistor to go into saturation

When the transistor do not saturate, the storage time will be reduced resulting in quick change from ON to OFF.

4.6 NON-SATURATING BINARY

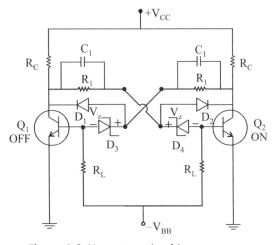

Figure 4.6 Non-saturation binary

The transistors are driven into saturation because the speed-up operation reduced and storage time delayed. The speed of operation can be increased by not allowing the transistor to go into saturation. Such binary in which transistor always operate in the active region only is called a non-saturating binary.

The diodes D_1 and D_2 are junction diodes while the diodes D_3 and D_4 are zener diodes. The zener diodes are always operational in breakdown region with zener voltage V across them with the polarities as shown. The V_2 is less than V_{cc}; the voltage across of the diodes D_1 and D_2 are very small because they are forward biased direction. When Q_2 is ON condition the emitter base junction is forward biased and $V_{BE} \square$ 0 V. The V_z drop across D_4 makes the left end of D_2 positive by V_z with respect to the ground. The right end of D_2 is at less voltage as collector voltage of ON transistor Q_2 is less than D_2 is forward biased. Thus the positive voltage V_2 reaches to the collector which makes $V_C B_2 = V_z$ this reverse biased of collector to base junction of Q_2 and do not allow Q_2 to enter into saturation.

$$V_R E_1 = -V_{BB}\left[\frac{R_1}{R_1 + R_2}\right]$$

The negative voltage keeps Q_1 OFF condition (cut-off). When Q_1 is OFF condition, V_{CE} is high; hence the diode D_1 is reverse biased. The output swing is approximately equal to $V_{cc} - V_z$.

When the very fast operation is required, then only non-saturation binary is used because of the following drawbacks.

1. It is more complicated than saturated binary
2. It consumes more power than saturated binary
3. The voltage swing is less stable with temperature agency and component replacement than in the case of saturated binary.

4.7 UNSYMMETRICAL TRIGGERING OF BISTABLE MULTI-VIBRATOR

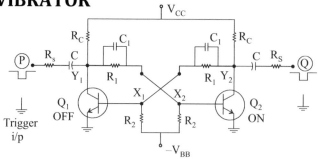

Figure 4.7 unsymmetrical triggering through a resistance and a capacitor **(a)** at the collector and **(b)** at the bases

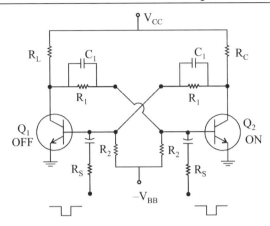

Bistable multi-vibrator has two stable states – the process of applying an external signal to induce a transition from one state to another state is called triggering. The triggering signal employed is either pulse of short duration (or) step voltage. There are two methods of triggering unsymmetrical triggered and symmetrical triggering.

The unsymmetrical triggering is one in which the triggering signal is affected in inducing a transistor in only one direction and the second triggering signal from separate source

In unsymmetrical trigger two trigger inputs are used one to set in particular stable state and other to reset the circuit to the opposite state.

This type of triggering signal is effective conducting in only one direction. When the signals are applied to the collector this measure Q_{ie} stable state ON condition Q_2 is OFF condition. Then triggering signal is applied to Q_1 input to change it OFF condition due to which Q_2 becomes ON condition. Then to change the state again, the triggering signal is then applied to input of Q_2 which is ON to make it OFF due to which Q_1 become ON. So two separates triggering signals are required to change the state of the multi-vibrator, every time changing the state either by changing ON transistor to OFF (or) OFF transition to ON. However in practice the unsymmetrical triggering is designed with two separate triggering sources to turn OFF the transistor which is ON. Such a triggering is also called set-reset triggering.

Let Q_2 be ON condition and Q_1 OFF condition and negative pulse is applied to the point a. i.e., of Q_1. The signal immediately opens at the base input ½ of the Q_2. These is due to the transition through commutating capacitor C_1 the present of series resistances R_s further increases the simulating of n-p-n transition to negative pulse. The Q_2 usually OFF condition of Q_2 and turn ON condition of Q_1.

It is necessary that is must be large in swing the C. To have next transition a second triggering signal is regulated at 1/2 point which quinsy appear at the base of Q_1. These turn OFF condition of Q_1 making Q_2 is ON condition so two separately triggering source are required to apply the is an unsymmetrical triggering.

An excellent method for triggering a binary in unsymmetrical on the leading edge of a pulse is to apply the pulse from high impedance source to the output of the non conducting device. For p-n-p transistor the positive pulse is required, if it is n-p-n transistor the negative pulse is required.

4.8 UNSYMMETRICAL TRIGGERING USING UNILATERAL DEVICE (DIODE)

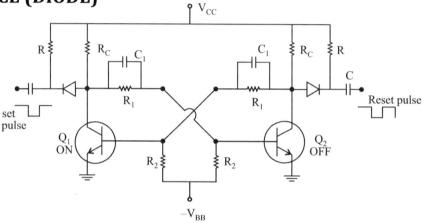

FIGURE 4.9 Unsymmetrical triggering at collector

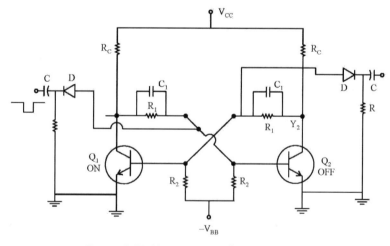

FIGURE 4.10 Unsymmetrical triggering at base

Fig (a) 4.9 shows unsymmetrical triggering through a unilateral device when the signal are applied at the collector. By using an unilateral device like diode un symmetrical triggering can be achieved which allows the bistable circuit to change the state onl because of the polarity of the pulse. Suppose in one stable state Q_1 is ON condition an

Q_2 is OFF condition. When Q_1 is ON condition. Accounting drop across R_c is large which keeps the diodes in reverse biased condition. Its anode is at V_{CE}(sat) and cathode is at V_{CC}. So, the diode will not transmit a positive pulse and even the negative cannot be transmitted unless the amplitude is large than the drop across R_c resulting in no change in the state of Q_2. So, no change of the state in the application of a pulse at the collector Q_1. When Q_1 ON condition and Q_2 is OFF condition both anode and cathode of diode collector are at V_{cc} and the drop across D is zero. The diode fails to transmit a positive triggering instead a negative triggering pulse (or) step input applying to the diode become forward biased condition and acts as a short circuited transmit the pulse to the base of Q_2 which is OFF makes it ON. Hence it is ensured that the transition occur at the leading as edge of the negative triggering pulse and the bistate circuit responds to that signal which effectively applies negative pulse to the ON device. R must be small enough so that any charge which accumulates on C during the interval when d conducts will have time to decay during the time between pulses.

Fig 4.10 shows that unsymmetrical triggering through a diode when triggering signal is applied to the base of the transistors. Negative pulse is applied through diode D is forward biased to the base the ON stage R returns to ground rather than to the supply voltage. While to active next transition a negative pulse is to be applied through reset terminal via diode D to the base of the Q_1 transistor.

4.9 TRIGGERRING SYMMETRICALLY THROUGH A UNILATERAL DEVICE

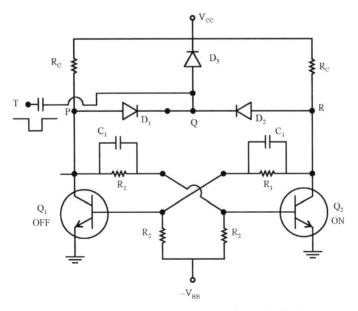

FIGURE 4.11 Symmetrical triggering through diodes

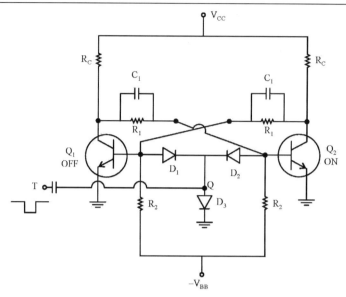

FIGURE 4.12 (a) at the collector **(b)** At the bases

In symmetrical triggering, the trigger input pulse is applied to only one transistor either base (or) collector.

Assume that if Q_2 is ON condition and Q_1 is off condition in one of the stable state the collector of Q_2 transistor is at VCE (sat) and the collector Q_1 transistor is at VCC. Hence p and Q are equal potential i.e., at VCC and drop across D_1 in zero this keeps diode D_1 a zero bias and D_2 in reverse biased.

Hence a negative going pulse is applied at the triggering input T. The point Q goe negative due to which diode D_1 gets forward biased condition. It act as short circuited hence pulse reached to part p is collector of Q_1 transistor. Then the negative pulse i passed to the base of Q_2 through the R_1 and C. The pulse switches OFF the transistor Q and Q_1 transistor become ON condition after the transition is completed D_1 will be reverse biased and D_2 will be zero bias.

In the consecutive negative triggering negative pulse is applied to point 'Q' goes negativ will pass through D_2 instead of through D_1. Hence these diodes are called steering diodes thus carried will not respond to the positive pulse.

If pnp transistors are used than diode direction must be reversed and positive pulse mus be used triggered.

Figure 4.12 shows as the arrangement at symmetrical triggering through the diode at the bases of transistor. When negative triggering pulse is applied to the base of Q_2 transisto through forward biased D_2 and Q_2 transistor become OFF condition. Q_1 transistor becom ON condition due to regenerative action. When successive negative triggering pulse i applied, D_1 become forward biased and pulse gets applied to the Q_1 which is then ON turning it OFF hence next transistor occurs.

.10 DIRECT-CONNECTION BINARY

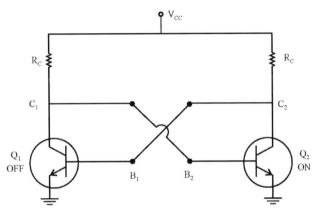

FIGURE 4.12 A direct connected binary

Figure shown a direct connect binary circuit. These binary circuit which does not used coupling elements is called direct connected binary. The collector of each transistor is connected to the base of the other transistor directly by accrue. In one stable state Q_2 is saturation than Q_1 is conditional slightly and not cut-off similarly another stable state Q_1 is saturation and Q_2 is conducting in Q_2 are

$$IB_2 = \frac{VCC - VBE_2}{RC} \text{ and } IC_2 = \frac{VCC - VBE_2}{RC}$$

direct to saturation, VBE_2 and VCE_2 are small compared to VCC. Thus $IB_2 \square IC_2$
$IB_2 >> IC_2/nfe$. Hence Q_2 is saturation in junction field.

Collector Q_2 is directly connected to base of Q_1 these $VBE_1 = VCE_2$ which is sufficient to make base-emitter junction of Q_1 slightly forward biased this Q_1 condition slightly.

Advantages:
1. Simple design
2. Dual supply not required supply of low value is sufficient about 15V
3. Less power departure
4. Transistor with low breakdown voltage rating can be used
5. Due to very few elements, it can be responded in integrated circuit (IC) man production

Disadvantages:
1. The output voltage switching is very small of the under of fraction of volt these saturation to remain voltage

2. The usual triggering methods can be used a trigger circuit with amplitude i necessary

3. As one of the transistor saturation, the storage time is more switching takes mor time

4. As temperature increases the resistance saturation circuit is I_{CBO} may increas sufficiently to binary Q_1 into active region and may even take Q_2 out of saturation

Application of Bistable multi-vibrator

1. Used in processing of pulse type waveform

2. Memory element in shift register, counter act

3. It perform many digital operations and storing of digital information

4. Generate the symmetrical square wave

5. Frequency deviser collector coupled

4.11   MONO STABLE MULTI-VIBRATOR

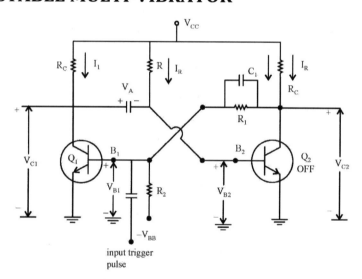

FIGURE 4.13 Mono stable multi-vibrator circuit

Fig 13 show the collector coupled mono stable multi-vibrator uses n-p-n transistor thi circuit is also called collector to base coupled mono stable multi-vibrator.

$$(Q_1 = \text{OFF} \quad Q_2 = \text{ON})$$

Q_1 and Q_2 transistor are n-p-n transistors, the collector Q_2 transistor is coupled to th base of Q_1 by a resistor R (dc coupling) and the collector of Q_1 is coupled to the bas of Q_2 by a capacitor C (ac coupling) C_1 is the cumulating capacitor. Introduced t increases the speed of operator. The capacitor to make the transistor fast and reduc the transistor time. The collector Q_1 is coupled to the base of Q_2 through a capacitor C The base of Q_1 is contained to-VBB through a resistance R_2 to ensure that Q_1 is cut-o under quiescent conductor. The base of Q_2 is connected to VCC through Resistance R to ensure that Q_2 is in ON condition. Q_1 is OFF and Q_2 transistor is ON normal stable state. Positive trigging is applied to the base of the Q_1 transistor capacitor C_2. When positive trigging of sufficient magnitude and duration is applied to the base of Q_1 transistor, the transistor Q_1 starts conducting due to the voltage at its collector VC_1 decreases coupled to base of Q_2 through C. but the voltage accrue capacitor (C) cannot change instantly. However decreases in VC_1 directly cause a decreases in the base of the voltage Q_2 i.e., V_{B2} the dropping voltage I_1R_C. Thus decreases of Q_2 and collector current I_2 decreases. The collector voltage at Q_2 is an increase which is applied to the base of Q_1 through R_1. These further the base polytonal of Q_1 quickly drive in to saturation condition at the same time. The transistor Q_2 gets into cut-off condition this is quiescent stable state of the circuit.

In the Quasi stable state the capacitor C starts charging through path V_{CC1}, R and transistor Q_1 is ON condition.

4.11.1 EXPRESSION FOR PULSE WIDTH OF COLLECTOR COUPLED MONO-STABLE MULTI-VIBRATOR

The pulse width is the time for which circuit remaining in the quasi state stable. It is also called gate width and denoted as T.

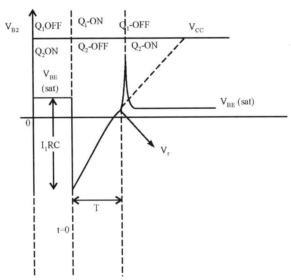

FIGURE 4.14 Voltage variation at the base of Q_2 during the Quasi stable state (ΔCBO neglecting)

For t < 0 Q_2 transistor is ON condition (saturation) and hence $V_{B2} = V_{BE}(sat) = V\sigma$ which is about 0.8 V for silicon transistor. When the pulse is applied at t = 0 then at t $= 0^+$ as capacitor voltage cannot change instantaneously, the voltage V_{BL} decreases by

$I_1 R_C$ V_{B2} as increases what increases exponentially and the capacitor change expositional when final value at $t = \alpha$ is V_{CC} V_{B2} equal to V_r then Q_2 starts conducting charges Q_1 is in OFF state exponentially.

Capacitor charging exponentially at equation $t = 0^+$ $V_i = V_o - I_1 R_c$

$$t = \alpha \qquad V_f = V_{CC}$$

$$V_e = V_f - (V_i - V_f)\, e^{-t/T}$$

$$T = R_c = \text{time constant}$$

$$VB_2 = V_{CC} - (V_{CC} - V\sigma + I_1 RC)\, e^{-t/\tau} \qquad\qquad .....(1$$

$$V_B = V_\gamma \qquad\qquad .....(2$$

1, 2 equation solving for T we get

$$T = T/n \left[\frac{V_{cc} + I_1 RC + V\sigma}{V_{cc} - V_\gamma} \right] \qquad\qquad .....(3$$

$$V\sigma = 0.3\ \text{V}$$

$$= 0.8\ \text{V}$$

When Q_1 is unsaturation under Quasi stable we can write

$$V_{C1} = V_{CE}\,(\text{sat})$$

$$I_1 RC = V_{CC} - V_{CE}(\text{sat})$$

Substituting in eq 3

$$T = T = \tau\Big|n \left[\frac{2V_{cc} - V_{CE}\,(\text{sat}) - V_{BE}\,(\text{sat})}{V_{cc} - V_\gamma} \right]$$

$$V\sigma = V_{BE}(\text{sat})$$

$$\tau\Big|n \left[\frac{2V_{cc} - V_{CE}\,(\text{sat}) - V_{BE}\,(\text{sat})}{V_{cc} - V_\gamma} \right] T$$

$$T = \tau\Big|n \left[\frac{2[\dfrac{V_{cc} - V_{CE}\,(\text{sat}) - V_{BE}\,(\text{sat})}{2}]}{V_{cc} - V_\gamma} \right]$$

$$T = \tau\Big|n + T\Big|n \left[\frac{[V_{cc}\dfrac{\{-V_{CE}\,(\text{sat}) - V_{BE}\,(\text{sat})\}}{2}]}{V_{cc} - V_\gamma} \right]$$

At this temperature $V_{CE}(\text{sat}) + V_{BE}(\text{sat}) = 2V_\gamma$

Substituting eq 4

$$T = T\big|n\ (2) + T\big|n\ (1)$$

$$T = T\big|n\ (2)$$

$$T = 0.69\ RC$$

The gate width T decreases as the temperature increases longer or more than value of V_{CC}.

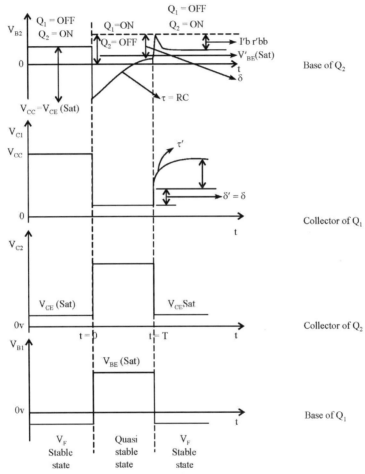

FIGURE 4.15 Waveform at the collector-coupled mono stable multi-vibrator

The waveform at the collector and base of both the transistor Q_1 and Q_2 of the mono stable multi-vibrator shown in Fig 4.15.

The Stable State: The mono stable multi-vibrator cut Q_1transistor is OFF condition and Q_2 transistor is ON condition when $t < 0$. Since the Q_2transistor ON and the base of voltage Q_2 is $VB_2 = VBE_2$ (sat), then collector voltage is $VC_2 = VCE_2$(sat). The

Q_1 transistor is OFF condition and no current flows the Rc of Q_1 and its base voltag may be negative. Hence the Q_1 transistor collector is

$$V_{C1} = V_{CC}, \qquad VB_2 = VBEsat = V\sigma$$

$$V_{C2} = V_{CE}(sat)$$

Q_1 transistor base voltage using super position theorem

$$V_{B1} = -V_{BB} \frac{R_1}{R_1 + R_L} + V_{CE2}sat \frac{R_2}{R_1 + R_2}$$

The Quasi stable state: When negative triggering signal is applied at t = 0, Q become OFF condition and Q_1 transistor become ON condition

The voltage at V_{C1} and V_{B2} drops instantaneously by the amount I_1RC where I_1 curren drown. Then by Q_1 when it starts conducting. Then Q_1 gets driven into saturation Hence

$$V_{B1} = V_{BE}(sat) = V\sigma$$

$$V_{C1} = V_{CE}(sat)$$

$$V_{B2} = V_{BE2}(sat) - I_1RC$$

$$V_{C2} = VCC \frac{R_1}{R_1 + RC} + V_{BE1}(sat) \frac{RC}{R_1 + RC}$$

The capacitor starts charging the voltage at base of Q2 raises exponentially a discussed earlier V_{CC}. This continues till V_{B2} become equal to cutting voltage V_γ at t = T

Waveform t > T: At t = T reverse transition occurs the Q_1 transistor is cut-of condition and Q_2 starts conducting.

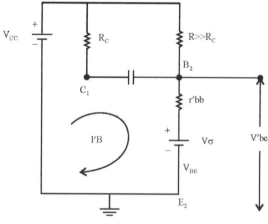

Figure 4.16 t > T

At t = T the V_{C2} drops instantaneously to V_{CE} sat and V_{B1} return to V_F. The V_{C1} voltage rises abruptly as Q_1 transistor becomes OFF condition. The rise is such that V_{C1} become almost equal to V_{CC}.

V_{C1} voltage sudden increases to applied to the base of Q_2 transistor and Q_2 transistor suddenly gets driven into saturation. Hence $t = T^+$. An overshoot occurs the at base of Q_2 is V_{B2} which decay as the capacitor recharge becomes the base current. The input circuit of Q_2 transistor is replaced by base spreading resistance r'bb in series with the base saturation voltage.

Let the base current I_B^+ at t = T while current through R is neglected as R >> R_e

$$V'_{BE} = I'b.r'bb + VBEsat(V\sigma)$$

and $\quad V_{C1} = V_{CC} - I_B' RC$

The jumps in voltage at B_2 and C_1 are respectively given by

$$\delta - V'_{BE} - V_r = I'b.r'bb + VBEsat - V_r$$

$$\delta' = VCC - VCEsat - I'b.RC$$

Since C_1 and B_2 are connected by a capacitor C and voltages across capacitor mostly dose not change instantaneously. These two discontinuos voltage change δ and δ' be equal.

Hence the overshoot δ and δ' must be same

$$I'b.r'bb + V\sigma - Vr = VCC - I'b.RC - VCE(sat)$$

$$I'B(RC + r'bb) = VCC - VCEsat - V\sigma + Vr$$

$$I'B = \frac{VCC - VCE(sat) - V\sigma + Vr}{Rc + r'bb}$$

After t = T^+, the VB_2 decreases exponentially to its steady state value $V\sigma$ the time constant with it decays is given by.

$$\tau' = (RC + r'bs)C$$

$$I'B \approx \frac{VCC}{RC} \quad \text{if VCC is large}$$

4.12   EMITTER COUPLED MONO STABLE MULTI-VIBRATOR

The figure 4.17 shows emitter coupled mono stable multi-vibrator. The emitter terminals of both the transistors are coupled together. Hence called emitter coupled. The connection from collector C_2 to the B_1 is absent. The base is provided through a common emitter resistance R_e. There is no need of negative power supply. Since the signal at C_2 is not directly involved in the regenerative loop.

This collector make an ideal point from where to obtain an output voltage waveform The input trigger is connected to terminal B_1 which is not connected to any othe point. This triggers input source cannot lead two circuits. The gate width at one sho can be controlled through I_1 but it is not possible to maintain I_1 stable but the emitte coupled one short the present of emitter resistance R_e. These serve to stabilize I_1 the current I_1 may be adjusted through the bias voltage V and T varies linearly with V When Q_2 becomes OFF as Q_1 transistor become ON condition and operate with sufficient emitter resistance.

The emitter coupled mono stable is used as a perfect gate wave generator. Gate width can be controlled easily and linearly with the help of a electrical signal.

Waveform: consider. The mode of operation Q_1 is cut off condition and Q_2 is saturation in stable state.

In the stable state Q_1 is OFF condition Q_2 is ON condition. Q_2 transistor is saturation i derive base from V_{CC} and resistance R. Due to it is emitter current produced a voltage VEN_2 across the resistance R_e. The voltage is more than the positional of Q_1 ensure that Q_1 remain in OFF condition

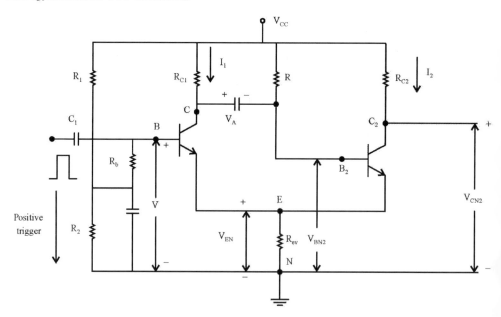

Figure 4.17 Emitter coupled mono stable multi-vibrator

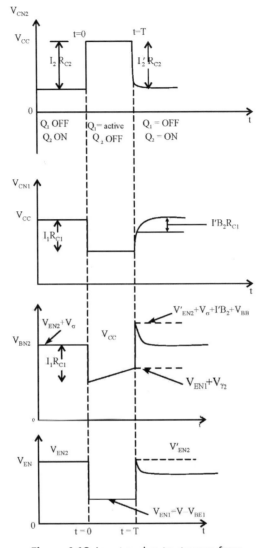

Figure 4.18 Input and output wave from

When a positive triggering pulse is applied at the base of Q_1 transistor which circuit goes into the Quasi – stable state becomes VB_1> VEN_2 whir drives Q_1 into conduction.

Due to this, the collector voltage of Q_1 transistor drop I_1 R_{CN} the negative step is applied to base of the Q_2 transistor which make Q_2 transistor in OFF condition. However Q_1 transistor is in ON condition. The capacitor C charges through R from VCC through ON transistor Q_1. The transistor Q_1 in ON condition develops potential drop of VEN_1 across RC. This is Quasi stable state with Q_1 in ON condition and Q_2 in OFF condition.

4.12.1 EXPRESSION FOR GATE WIDTH

The expression of gate width can be derived using the base relation.

$$VC = Vf + (Vi - V_f)e^{-t/T}$$

The VBN_2 voltage just after trigger applied is given by

$$VBN_2(0^+) = VBN_2(0^-) - I_1RC_1$$

Q_2 did not conduct ⋅ VBN_2 approach VCC.

$$VBN_2 = \text{instant voltage at } B_2$$

$$VBN_2 = VCC - [VCC - VBN_2(0^-) + I_1RC_1]e^{-t/\tau}$$

$$\tau = C[R+RC_1]$$

When $t = T^-, VBN2 = VEN_1 + V_{\gamma2}$

$$VBN_1 + V_{r2} = VCC - [VCC - VBN_2(0^-) + I_1RC_1]e^{-t/T}$$

Solving for T, the gate width is obtained

$$T = T/n \left\{ \frac{VCC - VBN_2(0^-) + I_1RC_1}{VCC - VBN_1 - V_{\gamma2}} \right\}$$

The time T varies linearly with the bias voltage V. Due to this feature it can be used a voltage to time converter.

4.13 TRIGGERING OF MONO STABLE MULTI-VIBRATOR

Triggering of mono stable is shown in fig 4.14

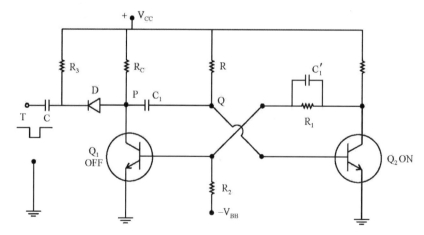

Figure 4.19 Triggering of mono stable multi-vibrator

The n p n transistor circuit negative pulse are saturated while for p n p transistor circuit. The positive pulse are selected negative pulse is applied through in capacitor with diode the n p n transistor. The more sensitive to the negative pulse. So as to make the ON and OFF transistor Q_1 transistor is OFF condition Q_2 transistor is ON condition are the normal stable state. The diode D zero biases at Q_1 is OFF condition, the P. Q points are equal potential. When the negative pulse is applied the diode D is forward biased (conducted) it act as a short circuit. This passes the negative pulse to the base of the Q_2 transistor is ON condition.

This decreases Q_2 transistor current and due to regenerated action of Q_1 gate drive to saturation and Q_2 transistor becomes OFF condition, this is Quasi-stable. The Quasi stable advantage is during due to drop across 'RC'. The diode is maintained reverse biased, during this state C_1 charge through R and Q_1, towards VCC. This base of voltage of Q_2 more than cut in voltage of Q_2 it becomes conducting. Hence due to the regenerative action automatically Q_1 becomes OFF conduction and Q_2 becomes coursed ON condition.

Application of Mono stable state multi-vibrator:

1. Gate width is adjustable

2. Introduces time delay

3. It produces rectangular waveform and hence can be used as gating circuit

4. Used to generate uniform width pulses from a variable width input pulse train.

5

Time Base Generators

5.1 ERRORS OF GENERATION OF SWEEP WAVEFORM

There are three most commonly used measure of sweep voltage. Sweep error displacement error and transmission error. These are called sweep parameters.

Sweep speed error (e_s)

It is important for sweep generator to keep sweep speed. Any change in sweep deviates sweep voltage from maintaining linear slope. The error due to sweep speed is called sweep speed error or slope error. It is given as

$$e_s = \frac{\text{Difference in slope at beginning and end of sweep}}{\text{Initial value of slope}} \qquad .....(5.1$$

5.2 DISPLACEMENT ERROR (E_D)

It is defined as the maximum difference between the actual sweep voltage and linear sweep which passes through the beginning and end points of the actual sweep as shown in fig 5.1(a)

$$e_d = \frac{(V_s - V'_s)_{max}}{V_s} \qquad .....(5.2$$

opened and the sweep voltage V_s is given by

$$V_s = V(1 - e^{-t/RC}) \qquad .....(5.3$$

If the switch is closed, after time interval, where is fig 5.1 Ts, when the sweep amplitude has attained the value v_s, the sweep waveform as shown in fig 5.1 (b)

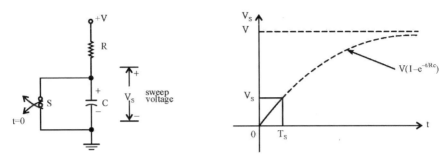

Figure 5.1 (a) exponential sweep circuit **(b)** Resultant exponential waveform

If switch resistance zero is assumed at o waveform is generated accordingly as shown in fig 5.5 (b) and here time interval T_r is zero.

Let the expressions for sweep speed error be derived (e_s), sweep displacement error (e_d) sweep transmission error (e_t) sweep speed for exponential charging and sweep speed for constant current charging.

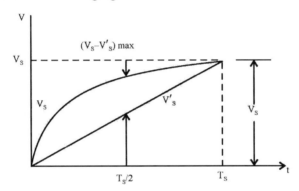

Figure 5.2 Related to displacement error

.3   TRANSMISSION ERROR

When a ramp voltage is transmitted through a high pass RC network its output falls away from the input as shown in fig 5.4. The transmission error is defined as the difference between the input and output divided by the input. Thus with reference to fig 5.3 we have

$$e_t = \frac{V'_s - V_s}{V'_s} \qquad\qquad .....(5.4)$$

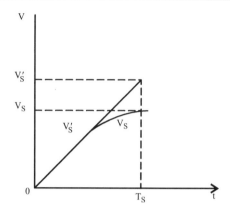

Figure 5.3 Related to transmission error

5.3.1 EXPONENTIAL SWEEP CIRCUIT

The fig 5.4 (a) shows the exponential sweep circuit. It consists of resistors capacito and switch connected across capacitor. At t = 0, the switch s is

Expression for Sweep Speed Error (e$_s$)

The sweep speed error is given as

$$es = \frac{\text{Difference in slope at beginning and end of sweep}}{\text{Initial value of slope}} \qquad .....(5.5$$

$$\frac{\left.\frac{dV_s}{dt}\right|_{i=0} - \left.\frac{dV_s}{dt}\right|_{t=Ts}}{\left.\frac{dV_s}{dt}\right|_{t=0}} \qquad .....(5.6$$

We know that exponentially increasing voltage can be given as

$$V_s = V(1 - e^{-t/RC}) \qquad .....(5.7$$

When RC = Time constant T

Differentiating equation (7) with respect to time,

$$\frac{dV_s}{dt} = V(0 - e^{-t/RC}) \times \left(-\frac{1}{RC}\right)$$

$$= \frac{V}{RC} e^{-t/RC}$$

$$\left.\frac{dV_s}{dt}\right|_{t=0} = \frac{V}{RC} \Rightarrow \frac{dV_s}{dt} = \frac{V}{RC} e^{-Ts/RC} \qquad .....(5.8$$

Substituting values from equation (8) in equation (6) we get

$$e_s = \frac{\dfrac{V}{RC} - \dfrac{V}{RC} e^{-Ts/RC}}{\dfrac{V}{RC}} \qquad \text{if Ts <<RC we write} \qquad \qquad .....(5.9)$$

$$e^{-Ts/RC} = 1 - \frac{T_s}{RC} = es = 1 - (1 - \frac{T_s}{RC}) = \frac{T_s}{RC} \qquad \qquad .....(5.10)$$

By definition sweep speed error is also given as

$$e_s = \frac{V_s}{V} \qquad \qquad .....(5.11)$$

where V is the supply voltage and Vs is sweep voltage

$$e_s = \frac{V_s}{V} = \frac{T_s}{RC} = \frac{T_s}{T} \qquad \qquad .....(5.12)$$

$\therefore$ Time constant $\tau = RC$

Expression for Displacement Error

The displacement error is given as

$$e_d = \frac{(V_s - V'_s)_{max}}{V_s} \qquad \qquad .....(5.13)$$

Exponential sweep is

$$V_s = V(1 - e^{-t/RC}) \qquad \qquad .....(5.14)$$

$1 - e^{-t/RC}$ can be written as

$$1 - e^{-t/RC} = 1 - \left(1 - \frac{t}{RC} + \frac{(t/RC)^2}{2!} - \frac{(t/RC)^3}{3!} + ...\right) \qquad .....(5.15)$$

Neglecting higher order terms in equation (15)

$$1 - e^{-t/RC} = 1 - \left(1 - \frac{t}{RC} + \frac{(t/RC)^2}{2}\right) = 1 - 1 + \frac{t}{RC} - \frac{(t/RC)^2}{2}$$

$$= \frac{t}{RC} - \frac{(t/RC)^2}{2}$$

$$= \frac{t}{RC}(1 - \frac{t}{2RC}) \qquad \qquad .....(5.16)$$

Substituting value of $1 - e^{-t/RC}$ in equation (14) is attained

$$V_s = \frac{Vt}{RC}(1 - \frac{t}{2RC}) \qquad \qquad .....(5.17)$$

The slope of linear sweep can be given as V/RC

$$V'_s = \frac{Vt}{RC}$$

$$\therefore \quad V_s - V'_s = \frac{Vt}{RC}(1 - \frac{t}{2RC}) - \frac{Vt}{RC}$$

$$= \frac{Vt}{RC} \times \frac{t}{2RC} \qquad .....(5.18)$$

Looking at fig 5.2 it is evident that deviation is maximum when $t = T_s/2$

$$\therefore \quad |V_s - V'_s|_{max} = \frac{V(T_s/2)}{RC} . \frac{(T_s/2)}{2RC} \qquad .....(5.19)$$

It is known that $V's = \frac{Vt}{RC}$ $\qquad .....(5.20)$

From fig 5.3 at $t = T_s$

$V'_s = V_s$ Substituting $t = T_s$ in equation (19) we have

$$V_s = \frac{VT_s}{RC}$$

Substituting values from equation (18) and (19) in equation (12)

$$e_d = \frac{\frac{V(T_s/2)}{RC} . \frac{(T_s/2)}{2RC}}{\frac{VT_s}{RC}}$$

$$e_d = \frac{(T_s/2)}{2RC} \qquad .....(5.21)$$

Expression for Transmission Error (e_t)

The transmission error is given as

$$et = \frac{V'_s - V_s}{V'_s} \qquad .....(5.22)$$

from equation (17)

$$V_s = \frac{Vt}{RC}(1 - \frac{t}{2RC})$$

The slope of linear sweep can be given as V/RC

$$V'_s = \frac{Vt}{RC}$$

At $t = T_s$, $V_s = V_s'$ and V'_s we have

$$V'_s = \frac{VT_s}{RC} \qquad .....(5.23)$$

And $V_s = \dfrac{VT_s}{RC}(1 - \dfrac{T_s}{2RC})$ .....(5.24)

Substituting values of V'_s and V_s from equations (23) and (24)

$$e_t = \dfrac{\dfrac{VT_s}{RC} - \dfrac{VT_s}{RC}(1 - \dfrac{T_s}{2RC})}{\dfrac{VT_s}{RC}} = 1 - (1 - \dfrac{T_s}{2RC})$$

$$et = \dfrac{T_s}{2RC}$$

Relation Between e_s, e_d, e_t

$$e_s = \dfrac{T_s}{RC}, e_d = \dfrac{T_s}{8RC} \text{ and } e_t = \dfrac{T_s}{2RC}$$

$$e_d = \dfrac{es}{8} \text{ (or) } e_d = \dfrac{et}{4}$$

Combining these equations

$$e_d = \dfrac{1}{8} e_s = \dfrac{1}{4} et$$

Because of this relation if one of the error is known other errors can easily be calculated.

Sweep Speed for Constant Current Charging

The sweep speed is defined as the rate of change of sweep voltage with respect to time. For sweep voltage for exponential charging. Instantaneous sweep voltage is given as

$$Vs = V(1 - e^{-t/RC})$$ .....(5.25)

Differentiating this expression with respect to time

$$\text{Sweep speed} = \dfrac{dV_s}{dt} = \dfrac{t}{RC} \times e^{-t/RC}$$ .....(5.26)

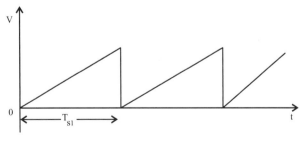

(a)

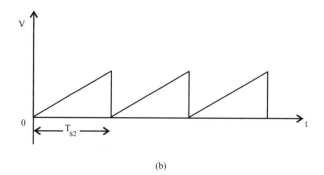

(b)

Figure (a) Low sweep speed (b) High sweep speed

Sweep Speed for Constant Current Charging

To generate linear sweep need to be charged capacitors with constant current. Charg on the capacitor is given as

$$Q = C\,V_C \qquad\qquad .....(5.27$$

where   Q is a charge on the capacitor

          C is the value of capacitor

          V_c is the voltage across capacitor

Differencing above equation with respect to 't' we get

$$\frac{dQ}{dt} = \frac{CdV_c}{dt}$$

$d\theta/dt$ is the charging current I of the capacitor

$$I = \frac{cdV_c}{\bullet\;dt} \qquad\qquad .....(5.28$$

$$\therefore\qquad \frac{dV_c}{dt} = \frac{I}{C}$$

Integrating both sides with respect to time 't'

$$\int\frac{dV_c}{dt}dt = \int\frac{I}{c}dt$$

$$V_c = \int\frac{I}{C}dt$$

In constant current charging I is constant for entire charging time

$$V_c = \frac{I}{c}t$$

and $\text{sweep speed} = \dfrac{dV_c}{dt} = \dfrac{dV_s}{dt} = \dfrac{I}{C}$

Table 5.1

Sl. No	Parameter	Expression
1.	Sweep speed error or slope error	$e_s = \dfrac{\text{Difference in slope at beginning and end of sweep}}{\text{Initial value of slope}} = \dfrac{T_s}{RC}$
2.	Displacement error	$e_d = \dfrac{(V_s - V'_s)_{max}}{V_s} = \dfrac{T_s}{8RC}$
3.	Transmission error	$e_t = \dfrac{V'_s - V_s}{V_s} = \dfrac{T_s}{2RC}$
4.	Sweep speed (exponential charging)	$\dfrac{dV_s}{dt} = \dfrac{t}{RC} \times e^{-t/RC}$
5.	Sweep speed constant current charging	$\dfrac{dV_s}{dt} = \dfrac{I}{C}$

5.4   UJT SAW TOOTH GENERATORS

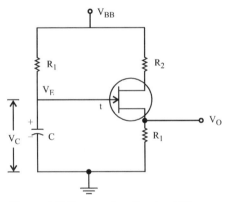

Figure 5.7 Sweep circuit using UJT

In the previous section the basic exponential sweep circuit in which mechanical switch is used in parallel with the capacitor C. Many devices such as transistor UJT, FET can be used as switch. In this section, impetus will be on how a UJT can be used as a switch to obtain the sweep voltage. The fig. 5.4 shows the sweep circuit using UJT.

Operation

Capacitor C gets charged through the resistor R towards supply voltage V_{BB}. As long as the capacitor voltage is less than peak voltage V_p the emitter appears as an open circuit.

$$V_p = \eta V_{BB} + V_D \qquad\qquad ..... (5.29)$$

where η = standoff ratio of UJT

V_D = cut in voltage of diode

When the capacitor voltage V_C exceeds the voltage V_p, the UJT fires. The capacitor starts discharging through $R_1 + R_{B1}$ where R_{B1} internal base resistance. As R_{B1} is assumed negligible and hence capacitor discharges through R_1.

The discharge time of the pulse is controlled by the time constant CR_1 while the charging time is controlled by the constant RC. The waveform is shown in the fig 5.8.

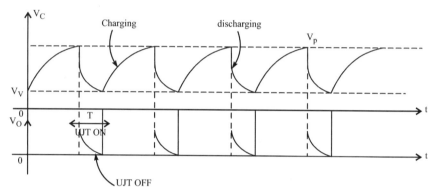

Figure 5.5 Wave form of sweep circuit using UJT

There is voltage drop across R_2 and voltage rise across R_1 when UJT the charging equation of the capacitor is given by

$$V_C(t) = V_v + V_{BB}\,[1 - e^{-t/RC}] \qquad .....(5.30)$$

But $V_C(t) = V_p{}'$ at tT

$$V_p = V_v + V_{BB}\,[1 - e^{-T/RC}] \qquad .....(5.31)$$

Substituting value of V_p from equation (5.29) in equation we have,

$$\eta V_{BB} + V_D = V_v + V_{BB}\,[1 - e^{-T/RC}] \qquad .....(5.32)$$

Neglecting V_v and V_D to get approximate relation for T

$$\eta = 1 - e^{-T/RC}$$

$$T = T_c = T_s = RC\,ln\,[\frac{1}{1-\eta}] \qquad .....(5.33)$$

$$To = \frac{1}{T} = \frac{1}{RC\,ln\,[\frac{1}{1-\eta}]} \qquad .....(5.34)$$

where = fo = oscillating frequency

Condition for Turn ON and Turn OFF

To ensure turn ON of UJT it should not be less than Ip at the peak point. To achieve this

$$V_{BB} - V_p > I_{pR}$$

$$R < \frac{V_{BB} - V_p}{I_p} \qquad \text{Turn ON condition}$$

To turn OFF the device, the valley point must be less than Iv specified thus voltage across R must be less than I_VR.

$$\therefore \quad V_{BB} - V_v < IV_R$$

$$R > \frac{V_{BB} - V_p}{I_p} \qquad \text{Turn OFF condition}$$

Hence the range of R for the proper turn ON and of OFF is

$$\frac{V_{BB} - V_p}{I_p} > R > \frac{V_{BB} - V_v}{I_v}$$

5.5 SEPARATE SUPPLY VOLTAGE OF UJT SAW TOOTH GENERATOR

For UJT sweep circuit $V_s = V_c - V_v$ from equation (12) from section 5.2 we have sweep speed error

$$Es = V_s/v$$

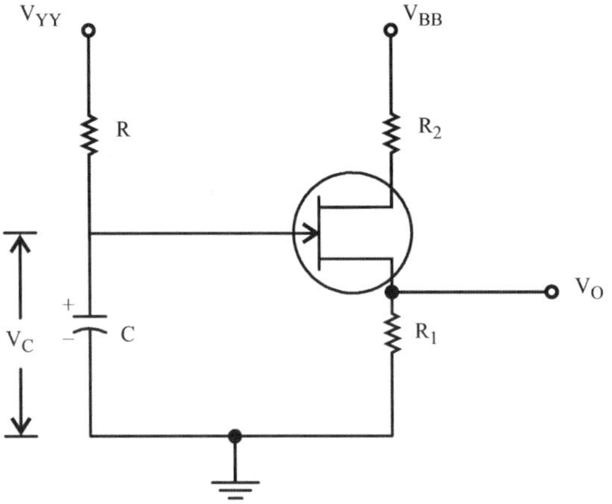

Fig 5.6 UJT sweep circuit with two supply voltages

Therefore, to have good linearity V_s must be much smaller than V. To achieve this in UJT sweep circuit $V_{BB} >> V_p$. However, for UJT circuit there is a restriction on the magnitude of V_{BB}. To solve this problem, two separate supplies are used as shown in fig 5.9 voltage. Supply V_{yy} is used for basic exponential circuit containing R and C and voltage supply V_{BB} is used for UJT circuit.

$$T_s = RC l n \left[\frac{V_{yy}}{V_{yy} - \eta V_{BB}} \right]$$ .....(5.35)

Other sweep parameters for UJT sweep generator are as shown in table 5.2

Table 5.2

Sl. No	Parameter	Expression
1.	Sweep speed error or slope error	$es = \dfrac{T_s}{RC}$
2.	Displacement error	$ed = \dfrac{T_s}{8RC}$
3.	Transmission error	$et = \dfrac{T_s}{2RC}$
4.	Sweep speed .	$\dfrac{dV_s}{dt} = \dfrac{t}{RC} \times e^{-t/RC}$

5.2 Expression for sweep parameter for sweep circuit using UJT

Figure

(b) Output waveform across the capacitor

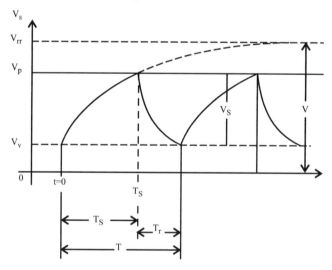

5.6 A TRANSISTOR CONSTANT CURRENT SWEEP

In common base configuration, the collector current of the transistor is almost constant with fixed emitter current except for very small values of collector-to-base voltage. This characteristic of common base configuration can be used to generate linear sweep by causing a constant current to flow into a capacitor. The fig 5.12 (a) shows the transistorised constant current seep generator.

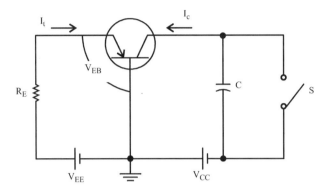

Figure 5.8 (a)

Looking at the figure 5.8 (a), the emitter current it can be given as

$$I_E = \frac{V_{EE} - V_{EB}}{R_e}$$ ..... (5.36)

If V_{EB} is assumed to remain constant with time after the switch S is opened, then the collector current will also be constant because $I_c = h_{fb}$; $I_E = - \alpha I_E$. As a result the capacitor C will charge linearly with time.

The fig 5.8 (b) shows h-parameter equivalent circuit. Here, the transistor is replaced by its CB hybrid parameter; sweep voltage is represented by V_s and the effective input signal V_i is given as $V_i = V_{EE} - V_r = V_r$ is the emitter threshold bias which brings the transistor just to the point of conduction.

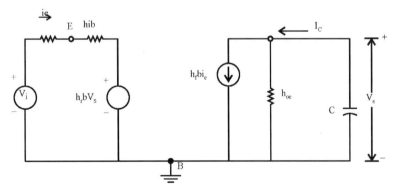

Figure 5.8 (b) Small signal equivalent circuit

By applying Kirchhoff's voltage law to the input circuit and Kirchhoff's current law to the output node in fig 5.8.

$$V_i = i_e (R_e + h_{fb}) + h_{rb} V_s \qquad \dots\dots(5.37)$$

$$i_C = i_e h_{fb} + h_{ob} V_s = -C \frac{dV_s}{dt} \qquad \dots\dots(5.38)$$

Considering initial conditions $V_s = 0$ at $t = 0$,

$$V_s = \frac{\alpha \tau V_i}{c(R_e + h_{ib})} (1 - e^{-t/T}) \qquad \dots\dots(5.39)$$

Where $\alpha = -h_{fb1}$, $Vi = V_{EE} - V_r$ and $\frac{1}{\tau} = \frac{1}{C} \left(h_{ob} + \frac{\alpha h_{rb}}{R_e + h_{ib}} \right)$

Expanding the exponential term into a power series in t/T and retaining only the first term

$$V_s = \frac{\alpha V_{it}}{C(R_e + h_{ib})} = \frac{\alpha i_{et}}{C}$$

$$\therefore \quad \frac{i}{C} = \frac{Vi}{(R_e + h_{ib})} \qquad \dots\dots(5.40)$$

$$= \frac{i_c t}{C} \quad \therefore i_c = \alpha i_c \qquad \dots\dots(5.41)$$

The sweep attitude V_s can be obtained by substituting $t = t_s$ in equation (6) as

$$V_s = \frac{\alpha ViT_s}{C(R_e + h_{ib})} \qquad \dots\dots(5.42)$$

$$T_s = \frac{V_s C(R_e + h_{ib})}{\alpha Vi} \qquad \dots\dots(5.43)$$

Slope error is given by

$$e_s = \frac{T_s}{T} = \frac{V_s C(RC + h_{ib})}{\alpha TVi} \qquad \dots\dots(5.44)$$

Substituting expression for T we have

$$e_s = \frac{V_s C(R_e + h_{ib})}{\alpha Vi} \times \frac{1}{C} \left(h_{ob} + \frac{-\alpha h_{rb}}{R_e + h_{ib}} \right)$$

$$e_s = \frac{V_s}{Vi} \left[\frac{h_{ob}(R_e + h_{ib})}{\alpha} + h_{rb} \right] \qquad \dots\dots(5.45)$$

5.7   MILLER VOLTAGE GENERATORS

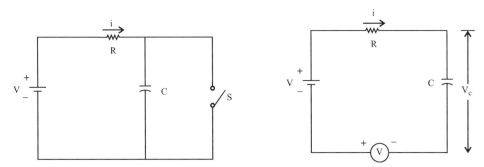

Figure 5.9 (a) Exponential charging **(b)** Constant current charging of capacitors

V and if V is always kept equal to the voltage drop across c, the charging current will be kept constant at i = V/r and perfect linearity can be achieved.

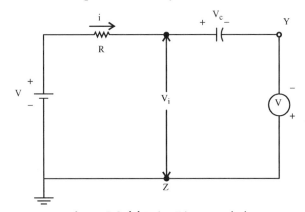

Figure 5.9 (c) Point Z is grounded

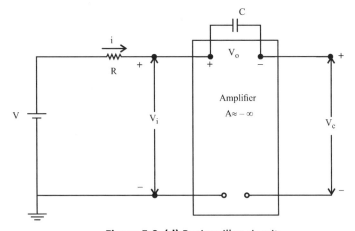

Figure 5.9 (d) Basic miller circuit

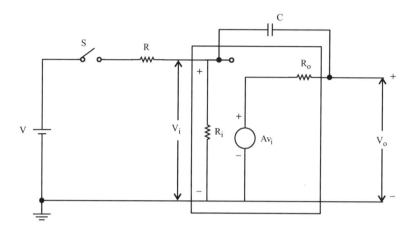

Figure 5.9 (e) Miller circuit with amplifier equivalent circuit

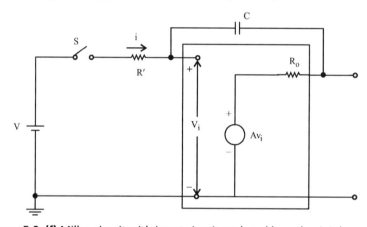

Figure 5.9 (f) Miller circuit with input circuit replaced by a thevinin's equivalent

$$V' = \frac{VR_i}{R_i + R} = \frac{V}{1 + R|R_i} \quad \text{and} \quad R' = \frac{R_i R}{R_i + R}$$

From fig 5.9 (f) let be assumed and at $t = 0^+$ voltage across capacious is zero; the neglecting R_o can be written

$$V_i - AV_i = V_i (1 - A) = 0$$
$$V_i = AV_i = 0$$

Since R_o is neglected $AV_i = V_o$

$$V_i = AV_i = V_o = 0$$

At $t = \infty$ when the capacitor is completely charged and no current flows through it

$$V_i = V' \text{ and}$$
$$V_o = AV' \qquad\qquad \therefore V_o = AV_i$$

Sweep Speed of Miller Circuit

From fig 5.8(c) charging is given as

$$i = \frac{V'}{R'} \qquad \qquad .....(5.46)$$

If a capacitor C is charged by a current i, then the voltage across C is $(i \times t)$ C. Hence, the rate of change of voltage with time is given by

$$\text{Sweep speed} = \frac{i}{c} \qquad \qquad .....(5.47)$$

Substituting value of i from equation (1) we get

$$\text{Sweep speed} = \frac{V'}{R'C} = \frac{V}{RC} \qquad \qquad .....(5.48)$$

Thus the sweep speed for Miller circuit is same as in the case where the capacitors charge through R directly from the source V.

5.8 BOOTSTRAP VOLTAGE GENERATORS

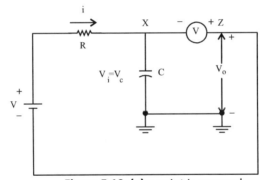

Figure 5.10 (a) y point is grouned

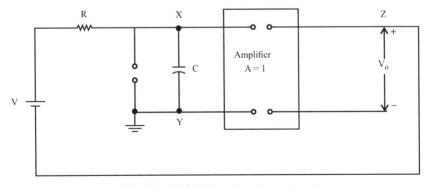

Figure 5.10 (b) Basic bootstrap circuit

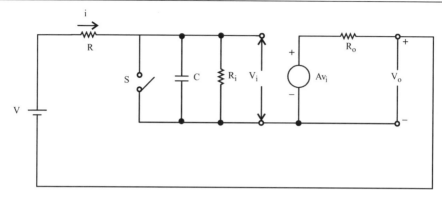

Figure 5.11 Basic circuit for bootstrap sweep generator

When S is closed,

$$V - iR + AVi - iRo = 0 \qquad .....(5.49)$$

At $t = 0^+$ voltage across capacitor is zero and hence

$$V_i = AVi = 0$$

∴   equation (1) becomes

$$V - iR - iRo = 0$$

$$V = i(R + Ro)$$

$$i = \frac{V}{R + R_0} \qquad .....(5.50)$$

The output voltage V_0 can be given as

$$V_0 = -iRo + AVi$$

$$= -iRo \qquad \because AVi = 0 \qquad .....(5.51)$$

Substituting value of i from equation (2)

$$V_0(at \ t = 0^+) = -\frac{V}{R + R_0} \times R_0 = \frac{-VR_0}{R + R_0} \qquad .....(5.52)$$

At $t = \infty$, current through C is zero and can be written as

$$i = \frac{V + V_0}{R + Ri} \ \text{and}$$

$$V_0 = -R_{oi} + AVi = -R_{oi} + ARii$$

Substituting the value of i

$$V_0 = -R_0\frac{(V + V_0)}{R + R_i} + ARi\frac{(V + V_0)}{R + R_i}$$

∴ $(R + Ri) V_0 + R_C Vo - AR_i V_0 = -RoV + ARi V$

∴ $[R + R_0 + (1 - A) Ri] V_0 = (-R_0 + ARi)V \Rightarrow V_0 = \frac{(-R_0 + ARi)V}{[R + R_0 + (1-A)R_i]}$

For emitter follower $A \approx 1$

$$\therefore \quad V_o = \frac{(-R_o + Ri)V}{R + R_o + (1-A)Ri}$$

For emitter followers $R_i >> R_o$ and $R >> R_o$

$$\therefore \quad V_o = \frac{RiV}{R + (1-A)Ri}$$

$$\therefore \quad V_o(\text{ât } t = 0) = \frac{V}{(1-A) + R/Ri} \qquad .....(5.53)$$

The slope error for bootstrap circuit can be determined as follows

$$e_s \text{ (boot strop)} = \frac{V_s}{V_o \text{at } (t = \infty) - V_o \text{at } (t = 0^+)}$$

$$= \frac{V_s}{V_o(\text{at } = \infty)} \qquad \therefore R_o << R_i \text{ we may neglect } V_o \text{ at } t = 0^+$$

Substituting value of at $t = \infty$

$$es \text{ (bootstrap)} = \frac{V_s(1 - A + R/R_i)}{V}$$

$$\text{sweep speed} = \frac{Ai}{C} = \frac{Av}{RC}$$

.9   PRACTICAL BOOTSTRAP SWEEP GENERATOR

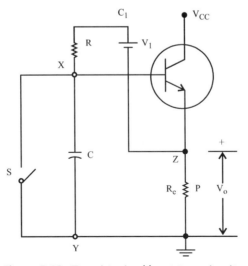

Figure 5.12 Transistorized bootstrap circuit

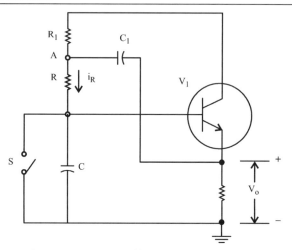

Figure 5.13 Practical boot strap sweep circuit

Transistorized Bootstrap Sweep Generator

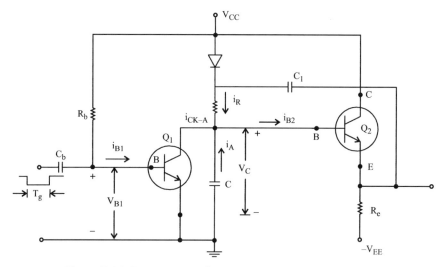

Figure 5.14 Boost strap suleys generats using transsion

Boot strap circuit with resistor R_1 replaced by D and switch S replaced by Q_1

$i_C = V_{EE}/Re$ the base current Q_2 for $t < 01$, $V_{EE} | R_e h_{fe}$

$$i_{C1} \approx iR = \frac{VCC}{R}$$

In order to keep Q_1 in saturation i_{B1} should be greater than $i_{C1} | h_{fcmin}$

$$\frac{VCC}{R_b} > \frac{VCC}{R_{hfe}} \text{ or } R_b < hfe\ R$$

Formation of Sweep

$$V_o = \frac{VCC^t}{RC}$$

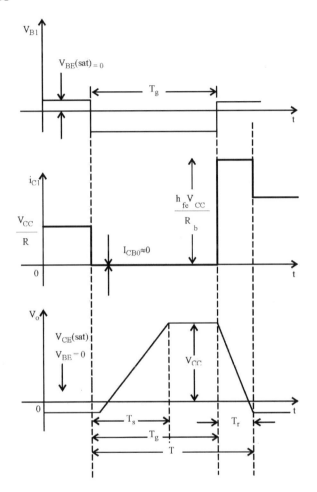

Figure 5.15 (a) Base voltage of Q_1 **(b)** Collector current of Q_1 **(c)** Output voltage

From the $t < 0$ Q_1 is ON, at $t = 0$ Q_1 is driven OFF as a result current i_{C1} flow into C_1 and assuming unity gain in the emitter follower, the output increases with time

$$V_o = \frac{VCC^t}{RC}$$

When V_o approaches, V_{CC1} the V_{CE} of Q_2 approaches and Q_2 enters in saturation. $T_s < T_g$ the T_s can be given as follows

$$V_o = \frac{V_{CC}T_s}{RC}$$

$$\therefore \quad T_s \approx RC$$

If sweep amplitude V_s is less than V_{CC}, then the maximum ramp voltage is given as

$$V_s = \frac{V_{CC} T_g}{RC}$$

Retrace Interval

$$i_{C1} = \frac{h_{fe} V_{CC}}{R_b}$$

If we apply KCL at point A

$$i_{C1} + i_{B2} = iR + iA \qquad\qquad iA = \text{discharge current}$$

Neglecting i_{B2} we get

$$i_{C1} = iR + iA \qquad \text{or} \quad iA = i_{C1} - iR$$

Substituting values of i_{C1} and iR

$$iA = \frac{h_{fe} V_{CC}}{R_b} - \frac{V_{CC}}{R}$$

$$= V_{CC}\left(\frac{h_{fe}}{Rb} - \frac{1}{R}\right)$$

i_{A1}, discharge current through C is constant, the voltage across C falls linearly with time is written as

$$V_S = \frac{iAT_r}{C}$$

$$T_r = \frac{V_s C}{iA}$$

Then $\quad T_r = \dfrac{V_s C}{V_{CC}\left(\dfrac{h_{fe}}{Rb} - \dfrac{1}{R}\right)}$

Recovery Process

In the recovery process the capacitor C_1 has to be charged to its initial voltage ($\approx V_{CC}$) At $t = T$ the diode D start conducting.

$$\text{Charge lost} = V_{CC}T$$

$$\text{Charge gained} = \frac{V_{EE} T_1}{Re} \qquad\qquad T_1 = \text{time required for recharging}$$

$$\therefore \quad \frac{V_{CC}T}{R} = \frac{V_{EE} T_1}{Re}$$

$$T_1 = \frac{V_{CC}}{V_{EE}} \frac{Re}{R}$$

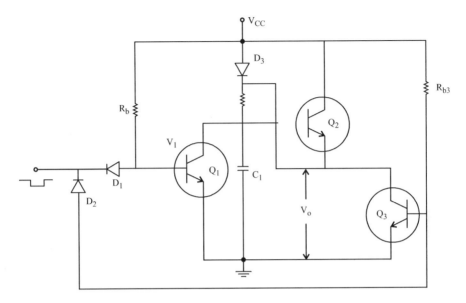

Q_3 used in bootstrap time base circuit for rapid recharge of C_1

Current Time Base Generators

In voltage time base generators, the voltage varies linearly with time. If such a voltage is applied to across resistance, the current I through the resistance will also vary linearly with time. The circuit which provides a current which linearly increase with time called current time base generators.

CHAPTER

6

Logic Gates

6.1   THE AND GATE

An AND gate has two or more inputs but only one output. AND gate is also called a
all or nothing gate.

Bellow figure shows symbols and truth tables of two-input and their input AND gates
Note that the output is 1 only when each one of the inputs is 1. The symbol for th
AND operation is .or we use symbol at all. The AND operation is logica
multiplication.

Truth Table		
Inputs		Output
A	B	X
0	0	0
0	1	0
1	0	0
1	1	1

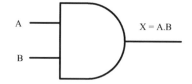

Figure 6.1 Logic Symbol

6.1 (a) Two-input AND gate

With the input variables to the AND gate represented by, A, B, C...., the Boolea
expression for the output can be written as X = A.B and lead as "X is equals to A an
B".

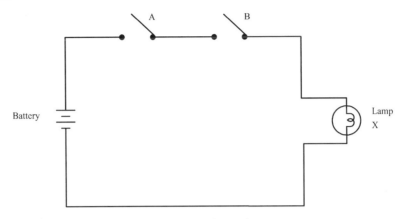

Figure 6.2 Operation of two i/p AND gate

A and B are mechanical switches connected in series and X is the lamp. It can be observed that the lamp X lights up only when both the switches A and B are closed. It can also be observed that the lamp does not light up with (a) both A and B open, (b) A open and B close and (c) A closed and B open. Let binary 0 indicate an open switch and binary one indicate a closed switch. Also let binary 0 indicate a dark lamp and binary 1 indicate a bright lamp. The various combinations of the switch positions and the state of the lamp are listed below.

Switches		Lamp		Switches		Lamp
A	B	X		A	B	X
Open	Open	Dark		0	0	0
Open	Close	Dark		0	1	0
Close	Open	Dark		1	0	0
Close	Close	Bright		1	1	1

This is the same as the truth table of a two-input AND gate.

6.2   THE NOT GATE (INVERTER)

A NOT gate also called as an Inverter. It is a device whose output is always the complement of its input.

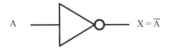

Figure 6.3 Logic symbol

Truth	Table
A	X
0	1
1	0

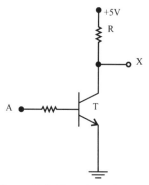

Figure 6.4 Transistor inverter

REALIZATION OF NOT GATE

(a) When A = 0 V, transistor T is OFF, NO current flows through R and hence no voltage drop occurs across R. The output voltage X = +5 V.

(b) When A = +5 V, T is ON. Current flows through R and hence almost all the supply voltage is dropped across R.

The output voltage $x = V_{CE}(sat) = 0.3$ V $\approx$ O V

The truth table for the NOT gate is shown below

Input	Output
A	X
Low	High
High	Low

Input	Output
A	X
0	1
1	0

The IC 7404 contains six inverters.

Logic circuits of any complexity can be realized using only AND, OR and NOT gates Logic circuits which use these three gates only are called AND/OR/INVERT i.e. AO logic circuits. Logic circuits which use AND gates and OR gates only are called AND/OR, i.e, AO logic circuits.

.3   THE UNIVERAL GATES

Though logic circuits of any complexity can be realized by using only the three basic gates, there are two universal gates, each one of which can also realise any digital circuit of any complexity single-handedly. The NAND and NOR gates are also, therefore, called universal building blocks. Both NAND and NOR gates can perform all the three basic logic functions. Therefore, AOI logic can be converted to NAND logic or NOR logic.

.3.1   THE NAND GATE

NAND means NOT AND i.e. the AND output is Noted. So, a NAND gate is essentially an AND gate and a NOT gate. The output of a NAND gate is therefore NOT the AND of the inputs. A NAND gate can have two or more inputs but only one output.

The output of a NAND gate is low only when all its inputs are high; its output is high even if one of its inputs is low.

Truth table

Inputs		Output
A	B	X
0	0	1
0	1	1
1	0	1
1	1	0

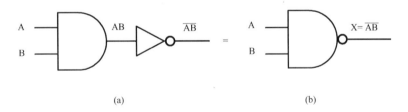

(a)   (b)

Figure 6.5 Two input N and gate symbol (a) (b)

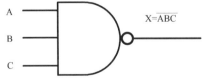

Figure 6.6 3 Input NAND gate symbol

Truth Table

Inputs			Inputs
A	B	C	X
0	0	0	1
0	0	1	1
0	1	0	1
0	1	1	1
1	0	0	1
1	0	1	1
1	1	0	1
1	1	1	0

With the input variables to NAND gate represented by A,B, C, ...the Boolean expression for the output X can be written as $X = \overline{AB}$ and read as "X is equal to AB whole bar" for a two-input gate. $X = \overline{ABC}$ and is read as "X is equal to ABC whole bar" for a three input gate.

Looking at the truth table of a two-input NAND gate, the that output X is 1 when either A = 0 or B = 0 or when both A and B are equal to 0. It is equal to zero only when both A and B are equal to I i.e., the output X is equal to 1 if either $\overline{A}=1$ or $\overline{B}=$ or both $\overline{A}$ and $\overline{B}$ are equal to 1. The output is 0 only when both $\overline{A}$ and $\overline{B}$ are equal to 0. Therefore, the NAND gate can perform the OR function. The corresponding output expression is, $X = \overline{A} + \overline{B}$. So, a NAND function can also be realized by first inverting the inputs and then OR ing the inverting inputs. Thus a NAND gate is a combination of two NOT gates and an OR gate. Hence, from, figure 6.7(a) we can express the output of a two-input NAND gate as

$$X = \overline{AB} = +\overline{A} + \overline{B}$$

Truth table

Inputs		Inverted i/ps		o/p
A	B	$\overline{A}$	$\overline{B}$	X
0	0	1	1	1
0	1	1	0	1
1	0	0	1	1
1	1	0	0	0

The OR gate with inverted inputs is called a bubbled OR gate. So, a NAND gate is equivalent to a bubbled OR gate whose truth is shown in figure 6.7(b). A bubbled OR gate is also called a negative OR gate. Since its output assumes the high state even if any one of its inputs is 0. The NAND gate is also called an active-LOW OR gate.

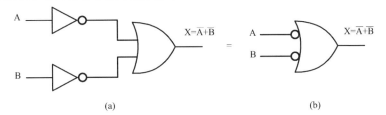

$X=\overline{A+B}$ = $X=\overline{A+B}$

(a)   (b)

Figure 6.7 Two input different NAND gate symbol (a, b)

6.3.2 THE NOR GATE

NOR means NOT OR, the OR output is NOTed. So, a NOR gate is essentially an OR gate followed by a NOT gate. A NOR gate can have two or more inputs but only one output. The output of the NOR gate assumes the state 1 only when each one of its inputs assume the 0 state, i.e. the output of the NOR gate is high only when all its inputs are low. Its outputs is low even if one of its inputs is high.

The logic symbols and truth tables of two input and three-input NOR gate are shown respectively.

Truth Table

Inputs		Output
A	B	X
0	0	1
0	1	0
1	0	0
1	1	0

Truth Table

Inputs			Output
A	B	C	X
0	0	0	1
0	0	1	0
0	1	0	0
0	1	1	0
1	0	0	0
1	0	1	0
1	1	0	0
1	1	1	0

With the input variables to NOR gate represented by A, B,C, ..the Boolean expression for the output X can be written as

$X = \overline{A + B}$ and read as "X is equal to A plus B whole bar" for a two-input gate.

$X = \overline{A + B + C}$, and read as "X is equal to A plus B whole bar" for a two-input gate.

Looking at the truth table of a three-input NOR gate, we see that the output X is 1 only when both A and B are equal to 0, i.e. only when both $\overline{A}$ and $\overline{B}$ are equal to 1, that means a NOR gate is equivalent to an AND gate with inverted inputs and the corresponding output expression is, $X = \overline{A} + \overline{B}$. So, a NOR function can also be realized by first inverting the inputs and then AND ing those inverted inputs. Thus, a NOR gate is a combination of two NOT gates and an AND gate. Hence, from and, we can see that the output of a two-input NOR gate is, $X = \overline{A + B} = \overline{A}\overline{B}$.

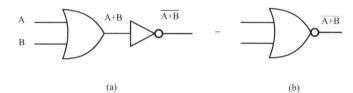

(a)   (b)

Figure 6.8 (a, b) Two input different NOR gate symbol

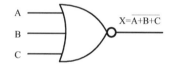

Figure 6.9 3 Input NOR gate symbol

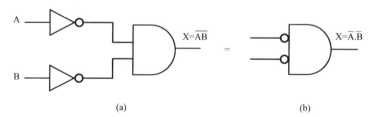

(a)   (b)

Figure 6.10 Two input different NAND gate

Truth Table

Inputs		Inverted inputs		Output
A	B	$\overline{A}$	$\overline{B}$	X
0	0	1	1	1
0	1	1	0	0
1	0	0	1	0
1	1	0	0	0

The AND gate with inverted inputs is called a bubbled AND gate. So, a NOR gate is equivalent to a bubbled AND gate whose truth table is shown in figure 6.10. A bubbled AND gate is also called a negative AND gate. Since its output assumes the HIGH state only when all its inputs are in LOW sates. A NOR gate is also called an active-low AND gate.

.4   THE DERIVED GATES

In addition to the three basic gates and two universal gates there are two more logic gates. They are the exclusive-OR (X–OR) gate and the exclusive-NOR (X-NOR) gate.

.4.1   THE EXCULSIVE-OR (X-OR) GATE

An X-OR gate is a two input, one output logic circuit, whose output assumes a logic 1 state when one and only one of its two inputs assumes a logic state. Under the conditions when both the inputs assume the logic 0 state, or when both the inputs assume the logic 1 state, the output assumes a logic 0 state.

The logic symbol and truth table of a two input X-OR gate are shown in figure 6.11, if the input variables are represented by A and B and the output variable by X, the expression for the output of this gate is written as $X = A \oplus B = A\bar{B} + \bar{A}B$ and read as "X is equal to A ex-or B".

Truth Table

Inputs		Output
A	B	$X = A \oplus B$
0	0	0
0	1	1
1	0	1
1	1	0

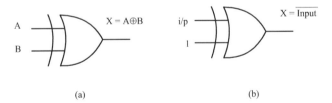

(a)   (b)

Figure 6.11 Two input different NOR gate symbol (a, b)

Three are more variable X-OR gates do not exist when more than two variables are to be X-o Red, a number of two input X-OR gates will be used. The X-OR of a number of variable assumes a 1 state only when an odd number of input variables assume a 1 state.

An X-OR gate can be used as an inverter by connecting one of the two input terminals to logic I and feeding the input sequence to be inverted to the other terminal as shown in figure 6.11(b). If the input bit is a 0, the output is, $0 \oplus 1$, and if the input bit is a 1, the output is, $1 \oplus 1 = 0$. In fact, an X-OR gate can be used as a controlled inverted, that is, one of its inputs can be used to decide whether the signal at the other input will be inverted or not.

The TTL IC 7486 contains four X-OR gates.

6.4.2 THE EXCULSIVE –NOR (X-NOR) GATE

An X-NOR gate is a combination of an X-OR gate and a not gate. The X-NOR gate is a two-input, one-output logic circuit, whose output assumes a 1 state only when both the inputs assume a 0 state or when both the inputs assume a 1 state. The output assumes a 0 state, when one of the inputs assumes a 0 state and the other a 1 state. It is also called a coincidence gate, because its output is 1 only when its inputs coinicide. It can be used as an equality detector because it outputs a 1 only when its inputs are equal.

The logic symbol and truth table of a two-input X-NOR gate are shown in figure respectively. If the input variables are represented by A and B and the output variable by X, the expression for the output of this gate is written as.

$$X = A.B = AB + \overline{A}\ \overline{B} = \overline{\overline{A} + B} = \overline{A + \overline{B}} + \overline{AB}$$

and read as "X is equal to A ex-norB".

(a)   (b)

Figure 6.12 NOR gate symbol of two input (a) (b)

Truth Table

Inputs		Output
A	B	X = A ⊙ B
0	0	1
0	1	0
1	0	0
1	1	1

The X-NOR of two variables A and B is the complement of the X-OR of those two variables.

That is,

$$A \Box B = \overline{A + B}$$

But the X-NOR of three variables A, B and C is not equal to the complement of the X-OR of A, B and C. that is,

$$A \Box B.C \neq \overline{A \oplus B \oplus C}$$

However, the X-NOR of a number of variables is equal to the complement of the X-OR of those variables only when the number of variables involved is even.

The TTL IC 74LS266, The CMOS IC 74C266 and the high speed CMOS IC 74HC266 contain four each X-NOR gates.

6.5 CMOS LOGIC FAMILIES

The basic buildings blocks in CMOS logic circuits are MOS transistors. Before going to discuss about MOS transitions and CMOS logic circuits. Let the logic levels in CMOs be dwelled.

6.5.1 CMOS LOGIC LEVELS

In general the CMOS circuit may interpret any voltage in the range 0 –1.5 V as logic and in the range 3.5 – 5.0 V as a logic1. As shown the voltages in between 1.5 V to 3.5 V are not expected to occur except during signal transitions and if they occur, the circuit may interpret them as either 0 to 1.

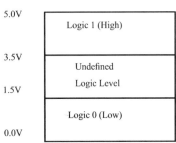

Figure 6.13 Logic levels for typical CMOS logic circuits

6.5.2 MOS TRANSISTORS

A MOS transistor is a three terminal device that acts like a voltage controlled resistance. An input voltage applied to one terminal controls the resistance between the remaining two terminals.

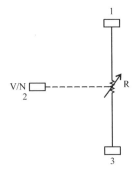

Figure 6.14 MOS transistor voltage contributed resolution

There are 2 types of MOS transistors: n-channel (NMOS) and p-channel (PMOS).

In NMOS transistor, the voltage from gate to source is normally zero or positive. I $V_{gs} = 0$ then the resistance from drain to source, Rd_s is very high. It is of the orde of few maga ohms. If v_{gs} is enough positive, then R_{ds} is very low. It is between 0 - 10 ohms.

The gate of the MOS transistor is separated from drain and source by an insulating material with a very high resistance. The voltage applied at the gate termina creates an electric field that enhances or retards the flow of current between sourc and drain. Due to this field effect the MOS transistor is also known as MOSFE (metal oxide semiconductor field effect) Transistor.

The small amount of current that flows across this resistance is very small typically less than one microampere and is called "A leakage current".

CMOS (complementary metal oxide semiconductor)

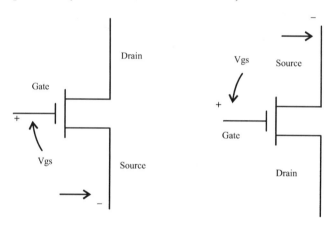

Figure 6.15 (a) NMOS Transition symbol **(b)** PMOS Transition, symbol

6.5.3 BASIC CMOS INVERTER CIRCUIT

It consists of two MOSFETs in series in such a way that the p-channel device has it source connected to $+V_{DD}$ (a positive voltage) and the N-channel device has it sourc connected to ground. The gates of the two devices are connected together as th common input and the drains are connected together as the common output.

1. When input is HIGH, the gate of Q_1 (p-channel) is at 0V relative to the sourc of Q_1 $V_{gs1} = 0V$. Thus Q_1 is OFF. On the other hand, the gate of Q_2 (N-channel is at $+V_{DD}$ relative to its source ie., $V_{gs2} = +V_{DD}$. Thus, Q_2 is ON. This wi produce $V_{out} \approx 0V$.

2. When input is low, the gate of Q_1 (p-channel) is at a negative potential relativ to it's source while Q_2 has $V_{gs} = 0V$. Thus Q_1 is ON and Q_2 is off. This produce output voltage approximately $+V_{DD}$.

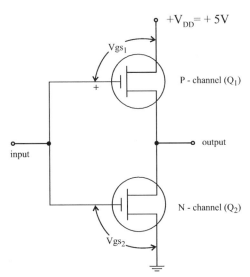

Figure 6.16 CMOS NOT gate

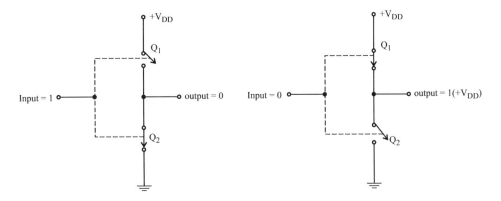

Figure 6.17 Operation of CMOS inverter for both input conditions

Truth Table of Invertor

A	Q_1	Q_2	Output
0	ON	OFF	1
1	OFF	ON	0

Different symbols used for the p-channel and n-channel transistors to reflect their logic behaviour. The n-channel transistor (Q_2) is switched 'ON' when a voltage is applied at the input. The p-channel transistor (Q_1) has the opposite behaviour. It is switched ON when a low voltage is applied at the input. It is indicated by placing bubble in the symbol.

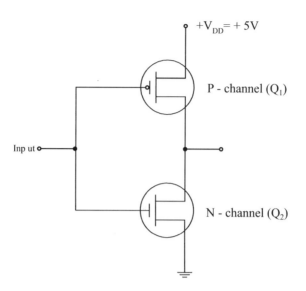

Figure 6.18 The CMOS inverter

6.5.4 CMOS NAND GATE

It consists of two p-channel MOSFETs, Q_1 and Q_2, connected in parallel and two N-channel MOSFETs, Q_3 and Q_4 connected in series.

It shows the equivalent switching circuit when both inputs are low. Here, the gates of both p-channel MOSFETs are negative with respect to their sources. Since the sources are connected to $+V_{DD}$.

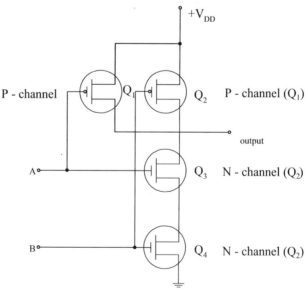

Figure 6.19 Two CMOS N gate using MOS transistors

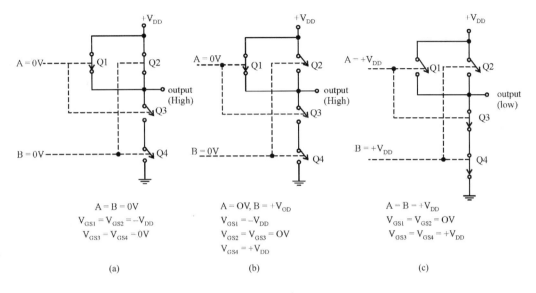

Figure 6.20 CMOS NAND gate operation (a, b, c)

The output is therefore connected to $+V_{DD}$ (HIGH) through Q_1 and Q_2 and is disconnected from ground, as shown in the fig (b) (c) shows the equivalent switching circuit when $A = 0$ and $B = +V_{DD}$. In this case Q_1 is on because $V_{GS1} = -V_{DD}$ and Q_4 is ON because $V_{GS4} = +V_{DD}$. MOSFETs Q_2 and Q_3 are OFF because their gate-to-source voltages are 0V. Since Q_1 is ON and Q_3 is OFF, the output is connected to $+V_{DD}$, and it is disconnected from ground.

- P-channel MOSFET is ON when its gate voltage is negative with respect to its source where as N-channel MOSFET is ON when its gate voltage is positive with respect to its source.

Truth Table of NAND gate

A	B	Q_1	Q_2	Q_3	Q_4	Output
0	0	ON	ON	OFF	OFF	1
0	1	ON	OFF	OFF	ON	1
1	0	OFF	ON	ON	OFF	1
1	1	OFF	OFF	ON	ON	0

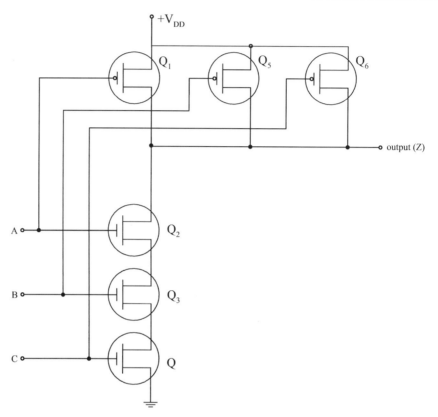

Figure 6.21 3 Input CMOS NAND gate using MOS transistors

A	B	C	Q_1	Q_2	Q_3	Q_4	Q_5	Q_6	Z
0	0	0	ON	OFF	OFF	OFF	ON	ON	1
0	0	1	ON	OFF	OFF	ON	ON	OFF	1
0	1	0	ON	OFF	ON	OFF	OFF	ON	1
0	1	1	ON	OFF	ON	ON	OFF	OFF	1
1	0	0	OFF	ON	OFF	OFF	ON	ON	1
1	0	1	OFF	ON	OFF	ON	ON	OFF	1
1	1	0	OFF	ON	ON	OFF	OFF	ON	1
1	1	1	OFF	ON	ON	ON	OFF	OFF	0

6.5.6 CMOS NOR GATE

Here, p-channel MOSFETs Q_1 and Q_2 are connected in series and n-channel MOSFETs Q_3 and Q_4 are connected in parallel.

Like NAND circuit, this circuit can be analyzed by realizing that not clear a low at any input ON it's corresponding p-channel MOSFET and turns OFF it's corresponding n-channel MOSFET, and viceversa for a HIGH input.

Truth table for NOR gate

A	B	Q₁	Q₂	Q₃	Q₄	Output
0	0	ON	ON	OFF	OFF	1
0	1	ON	OFF	OFF	ON	0
1	0	OFF	ON	ON	OFF	0
1	1	OFF	OFF	ON	ON	0

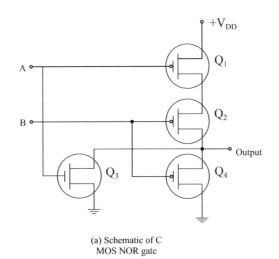

(a) Schematic of C
MOS NOR gate

Figure 6.22 (a) Schematic of CMOS NOT gate

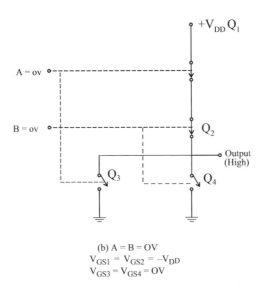

(b) A = B = OV
$V_{GS1} = V_{GS2} = -V_{DD}$
$V_{GS3} = V_{GS4} = OV$

Figure 6.22 (b) A = B = 0V

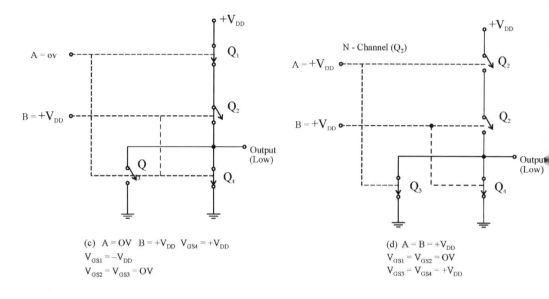

(c) $A = OV$ $B = +V_{DD}$ $V_{GS4} = +V_{DD}$
$V_{GS1} = -V_{DD}$
$V_{GS2} = V_{GS3} = OV$

(d) $A = B = +V_{DD}$
$V_{GS1} = V_{GS2} = OV$
$V_{GS3} = V_{GS4} = +V_{DD}$

Figure 6.22 (c) $A = 0V$, $B = + V_{DD}$ **Figure 6.22 (d)** $A = B = + V_{DD}$

Advantages and Disadvantages of CMOS family

Advantages

- Consumes less power
- Can be operated at high voltages resulting in improved noise immunity
- Fan-out is more
- Better noise margin

Disadvantages

- Susceptible to static charge
- Switching speed low
- Greater propagation delay

6.6   CHARACTERISTICS OF LOGIC FAMILY

There are various logic families. However, their nomenclature and terminologies use
by them are fairly standardized. The most useful terms are defined and discussed i
this section.

Fan out and fan in: In a digital system, we typically find many types of digital IC
inter connected to perform various functions. In these situations, the output of a logi
gate may be connected to the inputs of several other similar gates. The maximur
number of inputs of several gates that can be driven by output of a logic gate

decided by the parameter called fan-out. In general, the fan-out is defined as the maximum number of inputs of the same IC family that the gate can drive maintaining its output levels within the specified limits. For example, a logic gate with fan-out 10 can drive maximum 10 logic inputs from the same family. It depends on current sourcing and sinking capacity of input and output signals of same IC family.

6.7 TWO-INPUT TTL NAND GATE (STANDARD TTL)

In the circuit of the two-input TTLNAND gate shown in fig 9.1. The input transistor Q_1 is a multiple emitter transistor. Transistor Q_2 is called the phase splitter. Transistor Q_3 sits above Q_4 and, therefore, Q_3 and Q_4 make a totem pole arrangement. Diodes D_1 and D_2 protect Q_1 from being damaged by the Negative spikes of voltages at the inputs. When negative spikes appear at the input terminals, the diodes conduct and bypass the spikes to ground.

CIRCUIT DIAGRAM OF TTL

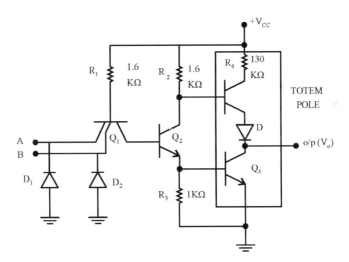

FIGURE 6.23

When both the inputs A and B are High (+5 V), both the base-emitter junctions of Q_1 are reverse biased. So, no current flows to the emitters of Q_1. The collector base junction of Q_1 is forward biased. So, current flows through R_1 to the base of Q_2 and Q_2 turns on; current from Q_2's emitter flows into the base of Q_4. So, Q_4 is turned on. The collector current of Q_2 flows through R_2 and, so, produces a drop across it thereby reducing the voltage at the collector of Q_2. Therefore, Q_3 is OFF. Since Q_4 is ON, Vo is at its low level (V_{CE} (sat)). So, the output is a logic 0. When either A or B or both are low, the corresponding base-emitter junctions. So, the current flows to ground through the emitters of Q_1. Therefore, the base of Q_1 is at 0.7 V, which cannot forward bias the base-emitter junction of Q_2. So, Q_2 is OFF with Q_2 OFF, Q_4 does not get the required base drive. so, Q_4 is also OFF. Transistor Q_3 gets enough base drive because Q_2 is OFF i.e, since no current flows into the collector of Q_2, all the current flows in to the base of Q_3, and therefore, Q_3 is ON.

The output voltage, $V_o = V_{CC} - V_{R2} - V_{BE3} - V_D \approx 3.4$ to 3.8V, which is a logic HIGH level. So, the circuit acts as a two-input NAND gate. When Q_4 is OFF, no current flows through it, but the stray and output capacitances between the output terminal i.e, the collector of Q_4, and ground get charged to this voltage of 3.4 to 3.8V.

6.8 TOTEM-POLE OUTPUT

In the circuit diagram of the two-input TTLNAND gate, transistor Q_3 sits above transistor Q_4.

Q_3 and Q_4 are connected in totem pole fashion. At any time, only one of them will be conducting. Both can't be ON or OFF simultaneously. Diode D ensures this. If Q_4 is ON, its base is at 0.7 V w.r.t ground. Q_4 gets base drive from Q_2. So, when Q_4 is ON Q_2 has to be ON. Therefore, it's collector-to-emitter voltage is V_{CE} (sat) ≈ 0.3 V Hence, $V_{B3} = V_{C2} \approx 0.7$ V + 0.3 V ≈ 1 V. For Q_3 to be ON, it's base-emitter junction must be forward biased. When Q_4 is ON, D has to be ON for Q_3 to be ON simultaneously. So, the base voltage of Q_3 must be $V_{B3} = V_{CE4}$ (sat) $+ V_D + V_{BE3} \approx 0.3$ + 0.3 + 0.7 ≈ 1.7 V, for it to be ON. Since V_{B3} is only 1 V when Q_3 cannot be ON Hence it can be concluded that Q_3 and Q_4 do not conduct simultaneously.

Advantages of totem-pole

1. Even though circuit can work with Q_3 and D removed and R_4 connected directly to the collector of Q_4, with Q_3 in the circuit, there is no current through R_4 in the output low state. So, the inclusion of Q_3 and D keeps the circuit power dissipation low.

2. In the output High state, Q_3 acts as an emitter follower with it's associated low output impedance. This action is commonly referred to as active pull up and it provides very fast rise time waveforms at TTL output.

Disadvantages of totem-pole

1. During transition of the output from low to High, Q_4 turns off more slowly than Q_3 turns on and so, there is a period of a few nano seconds during which both Q_3 and Q_4 are conducting and therefore, relatively large currents will be drawn from the supply.

2.   Totem-pole outputs cannot be wire ANDed, that is, the outputs of a number of gate can't be tied together to obtain AND operation of those outputs.

.9   OPEN-COLLECTOR GATES

The TTL gates may have totem-pole output or open-collector output. In open collector TTL, the output is at the collector of Q_4 with nothing connected to it. The open-collector inverter in order to get the proper HIGH and Low logic levels out of the circuit, an external pull-up resistor is connected to V_{CC} from the collector Q_4. When Q_4 is OFF, the output is pulled to V_{CC} through the external resistor.

➤   The open collector arrangement is much slower than the totem-pole arrangement because, the time constant with which the load capacitance charge in this case is larger.

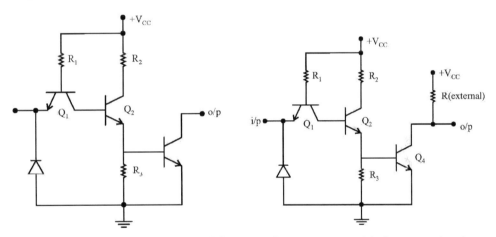

Figure 6.24 (a) Open-collector inverter **(b)** Open-collector inverter with the external resistor

TRI-STATE (3-STATE) TTL

The third TTL configuration is the tri-state configuration. It utilizes the advantages of high speed of operation of the totem-pole configuration and wire ANDing of the open-collector configuration. It is called the "tri-state TTL".

It allows three possible output states:

HIGH, LOW, and HIGH impedance (Hi-z). In this Hi-z state, both the transistors in the totem-pole arrangement are turned off, so, the input terminal is HIGH impedance to ground or V_{CC}.

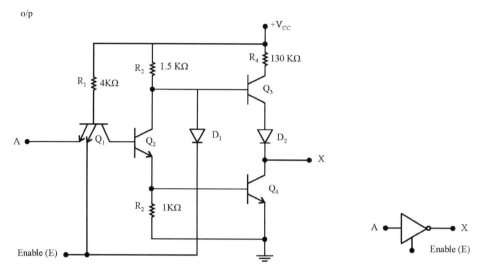

Figure 6.25 (a) Circuit diagram **(b)** Tri-state TTL inverter

CHAPTER

7

Sampling Gates

7.1 INTRODUCTION

An ideal sampling gate is basically a transmission circuit in which the output is an exact reproduction of an input waveform during a selected time interval and is zero otherwise. These gates are also referred to as transmission gates or as time-selection circuit. To get the exact replica of the input at the output of sampling gate during the selected time it is necessary that the input signal does not suffer any distortion or attenuation during transmission. In sampling gates the time interval for transmission is selected by an externally applied signal known as gating signal. The gating signal is usually rectangular in shape.

Sampling gates may be broadly classified as unidirectional sampling gates and bi-directional sampling gates. If the input signal consists essentially of a unidirectional pulse, the sampling gate is required to respond to an input signal of only one polarity, such sampling gate is known as uni-directional gate.

Sampling gates are different from logic gates, unlike, single input of sampling gate, the logic gate may have many inputs. The output of the logic gate is a voltage level or pulse which represents the logical combination of the input signals. It may not be exact replica of the input signal as in case of sampling gate during selected time interval; since in the sampling gate, the output waveform, is an exact replica of input waveform.

7.2 BASIC OPERATING PRINCIPLE OF SAMPLING GATES

The basic operating principle of sampling gates illustrated in fig (a) shows a circuit with a series switch in which the output is replica of input waveform when switch, S is closed on the other hand, in fig. 7.1 (b) the output is the replica of input waveform when switch, S is opened. The switch in fig 7.1 (b) is a shunt switch in fig (a) S is normally closed and is opened during the desired transmission interval.

For series switch: when S is closed $V_o = V_s$: otherwise $V_o = 0$

For shunt switch: when S is open $V_o = V_s$: otherwise $V_o = 0$

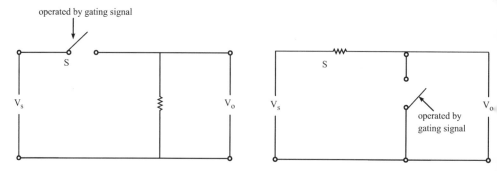

Figure 7.1 (a) Series switch **(b)** Shunt switch

- The inevitable stay capacitance across the switch will permit some signal transmission even when the switch is open.

- Since the signal is transmitted through S there will be some attenuation and distortion introduced by the nonlinearly of the device used for the switch.

Apart from the above disadvantages of series switch, the choice of switch for the circuit depends upon the particular application.

7.3   UNIDIRECTIONAL SAMPLING GATES

Unidirectional sampling gate which uses a semiconductor diode as a switch is shown in fig 7.2. This gate is suitable for a positive going input signal. The gate signal also called a control pulse, a selector pulse, or an enabling pulse is a rectangular waveform with voltages levels $-V_1$ and $-V_2$. As shown in figure 7.2, voltage level $-V_2$ is more positive.

The effect of the high level of gating signal $(-V_2)$ on the gate output is shown in fig 7.3 in which is considered as an ideal diode with $V_r = 0$ and amplitude of input signal pulse is super imposed on the gate signal. These two signals are differentiated by shading to different gray shades.

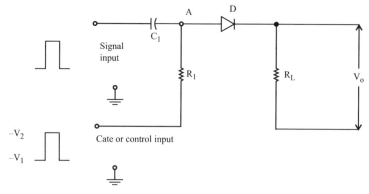

Figure 7.2 Uni-directional sampling gate

- When $-V_2 = -10$ V and $-V_1 = -20$ V, there is no output pulse at all as shown in fig 7.3 (a)

- When $-V_2 = -5$ V and $-V_1 = -20$ V, the output is $+10$ V pulse

- When $-V_2 = 0$ V and $-V_1 = -20$ V, the output is $+10$ V pulse

- When $-V_2 = +5$ V and $-V_1 = -20$ V, the output is a $+10$ V input pulse super imposed on a pedestal of $+5$ V

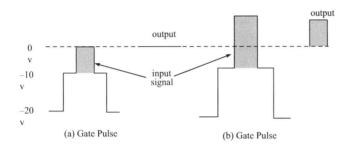

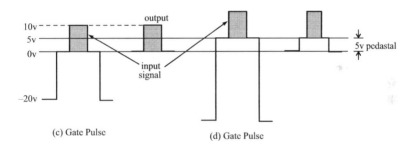

Figure 7.3 Illustrating the effect of control voltage $(-V_2)$ on gates o/p

Illustrating use of gate when gate width $\gg$ pulse width

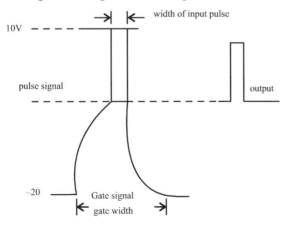

The advantage of unidirectional gate

1. It is very simple gate
2. The time delay is quite small, since the input is coupled directly to the output through C_1 and the node.

The disadvantages

1. There is undesirable interaction between the input signal source and gate voltage source.

7.4 UNIDIRECTIONAL SAMPLING GATES FOR MORE THAN ONE INPUT SIGNAL

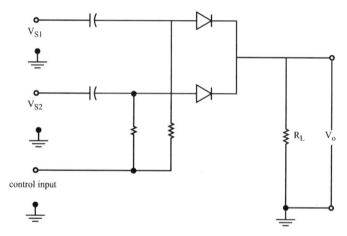

Figure 7.5 Unidirectional sampling gate with two input signals

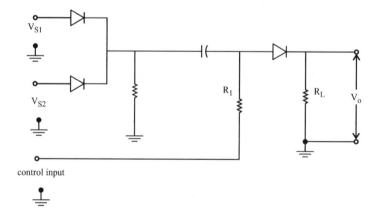

Figure 7.6 Sampling gate which avoids loading of gate signal

7.5   SAMPLING GATE WITH MULTIPLE GATE SIGNALS

The fig 7.7 shows the diode sampling gate with multiple gate signals. In this gate input is transmitted to the output only when all gate inputs are at their higher voltage levels. Do is removed and the input signal is transmitted at the output when any one of the gate signal V_c is at lower level i.e., at $-V_1$ point. A is negative with respect to ground.

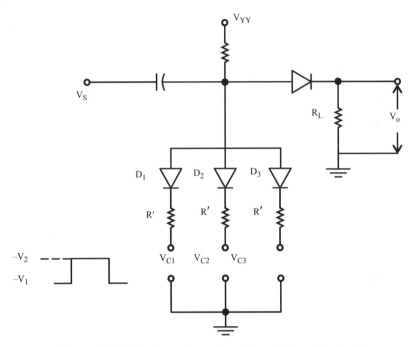

Figure 7.7 Diode Sampling gate with multiple gate signals

7.6   SAMPLING GATE NOT SENSITIVE TO HIGHER LEVEL OF GATE SIGNAL

When the gate signal is absent diode, D_1 conducts and the consequent voltage drop across R keeps diode Do back biased. Hence, there is no conduction and the output is o. On application of position going gate signal, diode D_1 gets reverse-biased and stops conduction. As a result diode Do gets forward-biased and input is transmitted to the output through the gate for the duration of the gate signal.

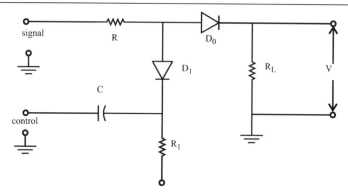

Figure 7.8 Diode sampling gate not sensitive to the output level of the gate voltage

7.7 BIDIRECTIONAL SAMPLING GATES

So for the focus of discussion is on unidirectional sampling gates. These gates have limitation that they pass only unidirectional signals. In this sections, we study the bi directional sampling gates. These gates can pass the signal of both polarities.

Bidirectional transistorized sampling gates are witnessed and the output wave forms of such sampling gates found the existence of pedestal in the wave form

Reduction of Pedestal in Gate Circuits

Bidirectional transistorized sampling gates. It in the observation is on output waveform of such sampling gates to find that existence of pedestal in the waveform.

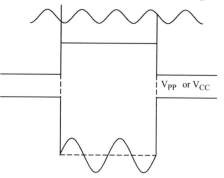

Figure 7.9 Illustrating the existence of pedestal

If the gate voltage has finite rise time then the circuit shown in fig 7.12 does no completely solve the problem of pedestal. This is illustrated in fig 7.13. It is assumed the gate bluer is large compared with active region

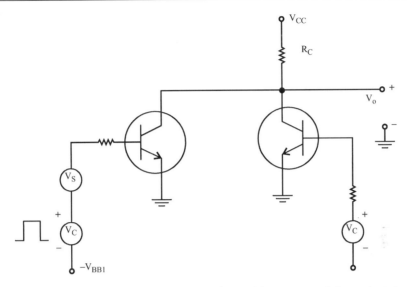

Figure 7.10 Linear gate circuit with provision to cancel the pedestal

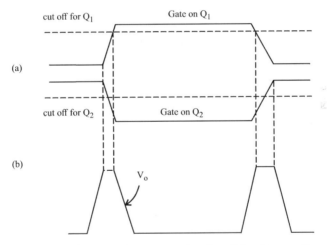

Figure 7.11 (a) Gating waveforms with nonzero rise time **(b)** voltage spikes at the output

.8 BIDIRECTIONAL DIODE SAMPLING GATE

It shows the bi-directional diode sampling gate using diodes instead of transistors. Such diode gates have two advantages.

- They ensure the linearity of operation and
- Ease of adjustment to get zero pedestal

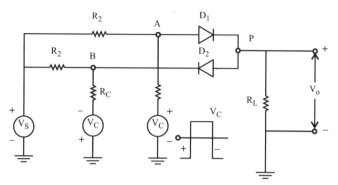

Figure 7.12 (a) Bi-directional sampling gate using diodes

The circuit is redrawn in the form of bridge. Two symmetrical gating voltages + V and $-V_c$ are now required. When the control signals are at the levels + V_n and – V respectively as observed from figure. The signal at A $-V_n$ and the signal at B is +V, So both the diodes D_1 and D_2 are reverse biased and hence no signal transmissio takes place.

$$\text{Gain } R_1 = R_2 \| RL = \frac{R_2 RL}{R_2 + RL} \text{, } R_3 = R_1 + R_{f1} \text{ } \alpha = \frac{RC}{R_2 + RL}$$

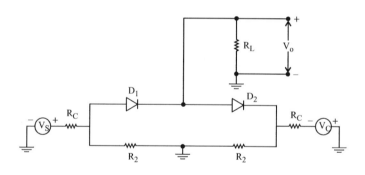

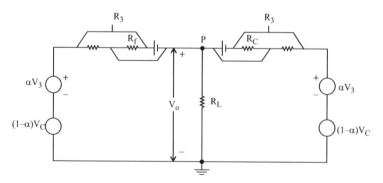

Figure 7.12 (b) (c) An equivalent circuit for the bidirectional diode gate

$$V_o = \alpha V_s \times \frac{RL}{RL + (R_3 / 2)}$$

So gain $\quad A = \dfrac{V_o}{V_s} = \alpha \dfrac{RL}{RL + (R_3 / 2)} \dfrac{RC}{RC + R_2} = \dfrac{RL}{RL + (R_3 / 2)}$

The Control Voltage V_c

$$\alpha V_s - (1 - \alpha)V_c$$

$$\therefore \alpha V_s - (1 - \alpha)V_c = \frac{RL}{R_3 + RL}[\alpha V_s + (1 - \alpha)V_c]$$

$$V_c = \frac{RC}{R_2} \frac{R_3}{R_3 + 2RL} V_s = (V_c)_{min}$$

The Control Voltage V_n

$$\alpha V_s - (1 - \alpha)V_n \qquad\qquad V_n \text{ is the magnitude of the Voltage}$$

$$\alpha V_s - (1 - \alpha)V_n = 0$$

$$V_n = \frac{\alpha}{1 - \alpha} V_s = \frac{RC}{R_3} V_s = (V_n)_{min}$$

.9   FOUR DIODE SAMPLING GATE

1.   Low gain
2.   It is sensitive to control voltage imbalance
3.   There is a possibility that $(V_n)_{min}$ may be excessive
4.   There may be appreciable leakage through the diode capacitances

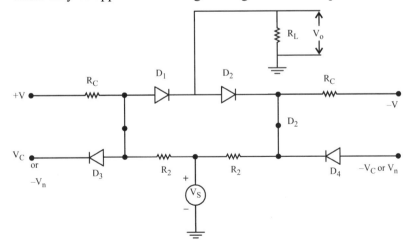

Figure 7.13 A four-diode gate

$$A = \frac{RC}{RC + R_2} \times \frac{RL}{RL + (R_3 / 2)}$$

Also the V_{min} of this circuit is the same as the $(V_c)_{min}$ of the gate of

$$V_{min} = \frac{RC}{R_2} \times \frac{R_3}{R_3 + 2RL} \times V_s \qquad \therefore (V_c)_{min} = AV_s$$

$$(V_n)_{min} = \frac{V_s RC}{RC + R_2} - V \frac{R_2}{R_2 + RC}$$

FOUR DIODE GATE (ALTERNATIVE FORM)

In these gate supply voltage $+V$ and $-V$ are not present but to avoid pedestal V_c and $-V_c$ must be balanced. The operation of the circuit is as follows.

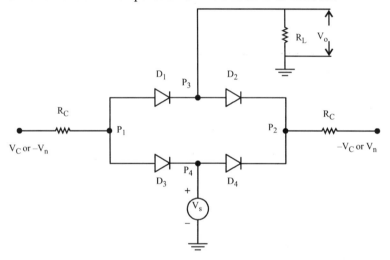

Figure 7.14 Diodes are replaced by shift circuits (a, b)

The required voltages V_c and $-V_c$ depend on the amplitude V_s of the signal and are determined by the condition that the current be in the forward direction in each of the diodes D_1, D_2, D_3 and D_4. Hence this quantity must be less than $V_c/2RC$. The minimum value of V_c occrue when

$$\frac{V_s}{RC} + \frac{V_s}{2RL} = \frac{V_c}{2RC}$$

$$(V_c)_{min} = V_s (2 + \frac{RC}{RL})$$

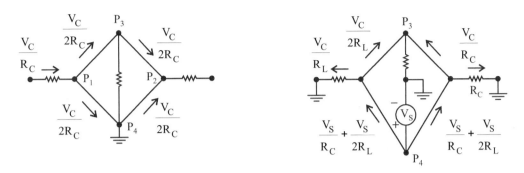

Figure 7.15 Diode is replaced by short circuits **(a)** Currents to V_c and **(b)** the current due to V_s

7.10   SIX DIODE SAMPLING GATE

The sampling gate using six diodes when output is zero. When there is no transmission, diodes D_5 and D_6 are forward biased and act as clamper, on the other hand, all other diodes are revere biased, puling transmission diodes D_1, D_2, D_3 and D_4 conduct while diodes D_5 and D_6 are reverse biased thus the diode gate which becomes equivalent to the four diodes as shown.

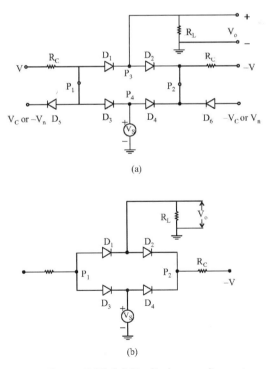

Figure 7.15 (a) Six diode sampling gate

According to equivalent, the current in each diode consists of two components one du...
to V and other due to V_s. The current due to $V = \dfrac{2}{2RC}$ and is in the forward direction i...
each diode. The largest reverse current is in D_3 and it is given by

$$ID_3 = \frac{V_s}{RC} + \frac{V_s}{2RL}$$

This current must be less than V/2RC i.e. , current due to V. Therefore the minimum
value V is given by

$$V_{(min)} = V_s \left(2 + \frac{RC}{RL}\right)$$

If the total resistance is R_1 and if R and R_f are $\ll$ RC or RL,

$$V_{(min)} = V_s \left(2 + \frac{RC}{RL}\right)\left(1 + \frac{R}{4R_f}\right)$$

As an example, if we assume $V_s = 20$ V, $R_f = 25\ \Omega$, RL = RC = 100 K, and R = 100 ...
then

$$V_{(min)} = 120\ V$$

Application of Sampling Gates

The importance of applications of sampling gates

1. Multiplexers
2. Sample and hold circuits
3. Digital to analog converters
4. Chopper stabilised amplifiers
5. Sampling scopes

Chopper stabilised amplifier and sampling scope as an applications sampling gates.

7.11 CHOPPER STABILISED AMPLIFIER

Suppose it is required to amplify a small signal and that signal V(t) is one in whic...
dv/dt is very small. Such signal cannot be amplified using ac amplifier because th...
required coupling capacitors will be impractically large. If we use dc amplifier for th...
amplification. The change in output voltage could not be distinguished as the chang...
in input voltages or as the result of a drift in some active device (or) componen...
voltages as the result of a change in input voltages or as the result of a drift in som...
active device or component.

The low frequency cut off ac amplifier that it passes high frequency square wav...
signal but blocks the low frequency input signal. As a result, we get only th...
modulated waveform at the output of this process of modulation, the chopper is ofte...
called a modulator.

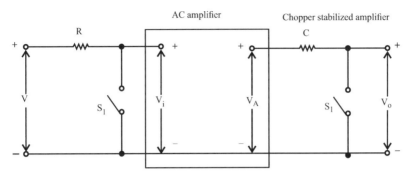

Figure 7.16 Chopper stateliest amplifier

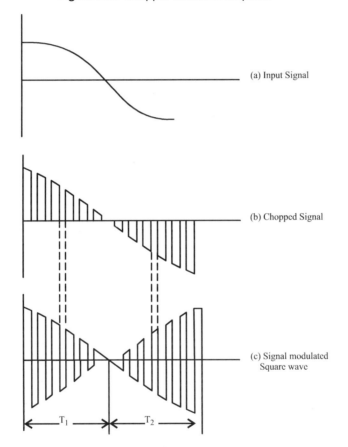

(a) Input Signal

(b) Chopped Signal

(c) Signal modulated
Square wave

Figure 7.17 (a) Input **(b)** Chopped signal **(c)** Signal modulated square wave

Sampling Scope

Sampling scope is an important application of sampling gates. It consists of a sequence of samples of the input waveform, each sample taken at a time progressively delayed with respect to some reference point in the waveform.

The diagram below shows the block diagrams of essential elements required for a sampling scope display. If needed charge the sentence.

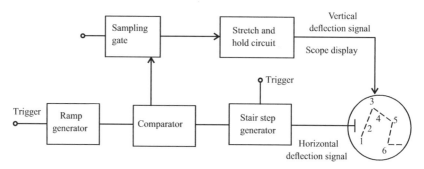

Figure 7.18 Block diagram of essential required for a sampling scope

Sampling gate. At each such gate signal, the sampling gate gives the sample of the signal as its output. The sample of signal as its output. The sample of input signal is kept so short that during its interval no sensible change takes place in the input signal. The gate signal holding operation may be performed. Sample in such a manner that when the sample is complained the capacitors holds its charge. If sampled signal is amplified before.

The stair step generator gives the horizontal deflection signal for the scope. Thus the CRT spot moves horizontally across the screen and at each new position.

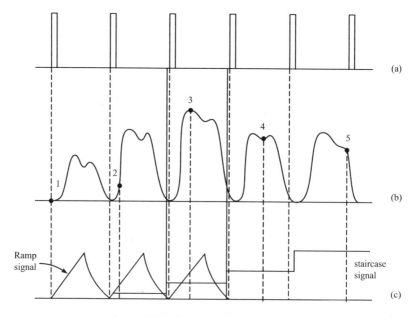

Figure 7.19 The sampling scope principle

(a) The triggering signal **(b)** The signal to be observed **(c)** The rump and stair step signal.

9781032228693